High Energy Electrons in Radiation Therapy

Edited by
A. Zuppinger J. P. Bataini
J. M. Irigaray F. Chu

With 46 Figures and 21 Tables

Springer-Verlag
Berlin Heidelberg New York 1980

ISBN-13:978-3-540-10188-8   e-ISBN-13:978-3-642-67727-4
DOI:10.1007/978-3-642-67727-4

The use of registered names, trademarks, etc. in the publication does not imply,
even in the absence of a specific statement, that such names are exempt from
the relevant protective laws and regulations and therefore free for general use.

Offsetprinting and Binding: Beltz Offsetdruck, Hemsbach/Bergstr.
2121/3140-543210

Preface

Radiotherapy using fast electrons, whether alone or in combination with high-voltage, has met with increasing interest in the last few years. This book provides a useful account of the present state of knowledge and critically discusses where an improvement of results is certain or probable — in contrast to results with radiotherapy using photons alone. The work also considers additional improvements which might be expected to accrue from past experience, and particular attention is paid to the nature and possible dangers of electron therapy.

Bern, August 1980   A. Zuppinger

Contents

Contents

List of Editors and Contributors

Altschuler, C.: Institut J. Paoli – I. Calmettes, 232, boulevard de Sainte Marguerite, F-13273 Marseille Cedex 2

Amalric, R.: Institut J. Paoli – I. Calmettes, 232, boulevard de Sainte Marguerite, F-13273 Marseille Cedex 2

v. Arx, A.: BBC Baden, CH-5400 Baden

Bamberg, M.: Radiologisches Zentrum, Universitätsklinikum der Gesamthochschule Essen, Hufelandstraße 55, D-4300 Essen 1

Bataini, J.P.: Institut Curie, Section Médicale et Hospitalière Radiotherapie, 26 rue d'Ulm, F-75231 Paris Cedex 05

Bernier, J.: Institut Curie, Section Médicale et Hospitalière Radiotherapie, 26 rue d'Ulm, F-75231 Paris Cedex 05

Botstein, C.: Montefiore Hospital and Medical Center, Department of Radiotherapy, 111 East 210th Street, USA-Bronx, NY 10467

Brahme, A.: Radiofysiska Institutionen, Karolinska Institutet, Box 60204, S-104 01 Stockholm

Brunin, F.: Institut Curie, Section Médicale et Hospitalière Radiotherapie, 26, rue d'Ulm, F-75231 Paris Cedex 05

Chu, F.: Department of Radiation Therapy, Memorial Sloan-Kettering Cancer Center, 1275 York-Avenue, USA-New York, NY 10021

Denepoux, R.: Fondation Bergonié Cancer Centre, 180 rue de Saint-Genès, F-33076 Bordeaux

Dobelbower, R.R., Jr.: Department of Radiation Oncology, Medical Physics, Medical College of Ohio, C.S. 10008, USA-Toledo, OH 43669

Eschwege, F.: Radiation Department, Institut Gustave Roussy, 16 bis avenue P.V. Couturier, F-94800 Villejuif

Greiner, R.: Klinik für Strahlentherapie, Inselspital, Universität Bern, CH-3010 Bern

Haie, C.: Radiation Department, Gustave-Roussy-Institute, 16 bis avenue P.V. Couturier, F-94800 Villejuif

Hünig, R.: Universitätsinstitut für medizinische Radiologie, Kantonsspital Basel, CH-4031 Basel

Ionesco-Farca, F.: Abteilung für medizinische Strahlenphysik, Inselspital, Universität Bern, CH-3010 Bern

Irigaray, J.M.: Instituto Oncologico de la Caja de Ahorros Provincia de Guipuzcoa, San Sebastian, Spanien

Jaulerry, C.: Institut Curie, Section Médicale et Hospitalière Radiotherapie, 26 rue d'Ulm, F-75231 Paris Cedex 05

Kalnicki, S.: Montefiore Hospital and Medical Center, Department of Radiation Therapy, 111 East 210th Street, USA-Bronx, NY 10467

Lagarde, C.: Fondation Bergonié Cancer Centre, 180 rue de Saint-Genès, F-33076 Bordeaux

Neiger, M.: Klinik und Poliklinik für Hals-Nasen-Ohrenleiden, Hals- und Gesichtschirurgie, Inselspital, Universität Bern, CH-3010 Bern

Paliwal, B.R.: Department of Human Oncology, University of Wisconsin, University Hospitals, 1300 University Avenue, USA-Madison, WI 53706

Paunier, J.P.: Centre de Radiothérapie Hôpital Cantonal, CH-1211 Genève 4

Pigneux, J.: Fondation Bergonié Cancer Centre, 180 rue de Saint-Genès, F-33076 Bordeaux

Poretti, G.: Abteilung für medizinische Strahlenphysik, Inselspital, Universität Bern, CH-3010 Bern

Pourquier, H.: Centre Paul Lamarque, 2, avenue Bertin Sans, Cliniques St. Eloi, F-34033 Montpellier Cedex

Richard, J.M.: Cervicofacial and Surgery Otorhinolaryngology Department, Gustave-Roussy-Institute, 16 bis avenue P.V. Couturier, F-94800 Villejuif

Richaud, P.: Fondation Bergonié Cancer Centre, 180 rue de Saint-Genès, F-33076 Bordeaux

Robert, F.: Institut J. Paoli – I. Calmettes, 232, boulevard de Sainte Marguerite, F-13273 Marseille Cedex 2

Rosenwald, J.C.: Institut Curie, Section Médicale et Hospitalière, 26 rue d'Ulm, F-75231 Paris Cedex 05

Scherer, E.: Radiologisches Zentrum, Universitätsklinikum der Gesamthochschule Essen, Hufelandstrasse 55, D-4300 Essen 1

Scholz, A.: BBC Baden, CH-5400 Baden

Spitalier, J.M.: Institut J. Paoli — I. Calmettes, 232, boulevard de Sainte Marguerite, F-13273 Marseille Cedex 2

Spittle, M.F.: The Meyerstein Institute of Radiotherapy, The Middlesex Hospital, St. John's Hospital, GB-London WIN 8AA

Strubler, K.A.: Department of Radiation Therapy and Nuclear Medicine, Thomas Jefferson University Hospital, USA-Philadelphia, PA 19107

Svensson, H.: Department of Radiation Physics, S-90185 Umeå

Touchard, J.: Fondation Bergonié Cancer Centre, 180 rue de Saint-Genès, F-33076 Bordeaux

Vaisman, I.: Department of Radiation Oncology, Institute Diagnostic, Caracas, Venezuela

van der Pol, P.C.: Municipal Hospital „Leyenburg", Leyweg 275, Den Haag, The Netherlands

Veraguth, P.: Klinik für Strahlentherapie, Inselspital, Universität Bern, CH-3010 Bern

Wibault, P.: Radiation Department, Gustave-Roussy-Institute, 16 avenue P.V. Couturier, F-94800 Villejuif

Wiley, A.L., Jr.: Departments of Human Oncology and Radiology, University of Wisconsin Center for Health Sciences, USA-Madison, WI 53706

Zuppinger, A.: Alpenstrasse 17, CH-3006 Bern

Opening Address

A. Zuppinger

It is a great honor and a pleasure for me to welcome here — in San Sebastian, at the symposium on the use of high energy electrons in radiotherapy — the Basque authorities, the director of the Caja de Ahorros provincial de Guipuzcoa Carlo Sistiaga, the president of the Spanish Radiological Society (Serem) Dr. Sanchez Pedroza, the medical director of the Oncological Institute Dr. Uriate, and all the members of the symposium. Doctor Irigaray, the chief radiologist of the Istituto Oncologico de la Caja de Ahorros Provincial de Guipuzcoa, is the initiator of this meeting. He received unique and generous help and assistance from the organizers of the Caja, especially from the director himself.

For more than 3 years they have worked here with a 45 MeV Asklepitron, and they found by experience that in several tumor situations the results achieved were better than with the former methods of radiotherapy. We are very obliged to them, and we would like to thank in advance the Oncological Center, the medial director, Dr. Uriate, and all those who have helped, for enabling us to exchange our experiences with electron therapy. I am convinced that we will hear important information in which clinical situations the application of electrons or a combination of electrons and photons will be advantageous for the patients, where the limits lie, which problems still exist, and how we can solve them in the safest and quickest way.

Fourteen years ago a symposium on high energy electrons took place in Montreux under the patronage of the European Society of Radiology. Basic physics, radiobiologic, and clinical experiences were discussed. In the meantime, the physical basis has become very much more reliable, so that we can now measure the absorbed doses with much greater accuracy, and we can calculate the doses at the place of the desired effect. As the papers of the physicists will show, many problems are still not solved satisfactorily. But the premises of good treatment planning are now much better than at the time of the Montreux Symposium. We know now, or we can predict, the most important tumor situations in which with electrons or electrons in combination with photons better results can be expected than with photons alone. This knowledge per se, its discussion, and its distribution in the medical sphere justify our present discussion.

But it has been known for 14 years that some early results could not be explained by the physical turnover of the energy in the crucial tissue, why some special tumors and tumor situations — which will be discussed later on — are controlled much better with electrons than with photons. These observations have been confirmed to a great extent, as we will hear in this symposium. The results obtained with electrons are probably but not certainly better, and I know that some of the participants of this symposium doubt my words on this point. In medicine, especially in the field of the battle against cancer, it often takes many years, even decades, until a medically probable effect can be confirmed or rejected. This fact is not caused by insufficient methods, but mainly by the fact that in biology and thus in medicine, and here in the special field of radiotherapy, it has been impossible to express many parameters quantitatively. The evaluation of what we really do is only possible in clinical experiments on a broad basis.

These have to be based and planned upon our former experiences and must strongly follow the principles of medical ethics. These ideas were already known at the time of our first symposium, but we left the performance to the initiative of the single radiotherapist. A further meeting 5 years later had been planned but it could not be realized. We certainly can make better and more reliable statements now, but they will probably not give us the necessary certainty. We thus have to change our philosophy. One has learned in the meantime to reduce the time interval necessary to answer the question of whether the results of our treatment really correspond to progress or not. One must try to join different centers into several distinct working programs. We may learn much from the work of the chemotherapists, who made many statistics and who succeeded in getting a good acknowledgment from the majority of the medical doctors, even though their results are in the great majority of patients only palliative. As we have in our field not only palliation, but very often a suprisingly good cure rate, we are not only justified, but I think even obliged, to proceed in a similar way. I would like to propose that we create working groups which allow us to judge earlier what results we can achieve with the different methods of electron therapy, also in combination with photons or even with chemotherapy.

Physical Section

Introduction

J.L. Mincholé

Applying radiotherapy by using accelerated electron beams between 3 and 45 MeV is quite well established now a days. Problems related to physics when using this type of irradiation have been basically solved, even though many of the practical aspects still have doubtful areas and justify an interchange of ideas and experiences among those who operate in this field as well as among those who construct the instruments which provide accelerated electrons.

The use of linear and circular accelerated electrons, introduced in Spain in 1966, was not generally accepted until these past few years. Considering this point, and because our own experience is very limited, we have to admit that we are thinking more of what we are going to learn than of what we will be able to offer. Consequently, we wish to express our deepest thanks to each and all the colleagues who are coming to share their experiences with us and thus help us improve many aspects of electron therapy.

Computer Treatment Planning of Lung Radiation by Means of High Energy Electrons*

G. Poretti, F. Ionesco-Farca, and P. Veraguth**

The dose distribution in the body of patients who are subjected to radiation with high energy electrons depends on the following known factors: focus skin distance (FSD), rotation angle, electron energy, curvature of the body surface, and shape and composition of the organs.

With the aid of the computer, which is capable of simulating a multiple-field or moving-beam irradiation by the summation of the dose values of two or more single fields, the above mentioned parameters can be changed as required, and thus various types of treatment plans can be demonstrated. Prior to the treatment, the radiotherapist can select the optimal plan for the case in question from amongst a number of possibilities. This study will report on the computer-assisted determination of isodose plans of lung radiation by means of high energy electrons.

General Procedure

The dose distribution in the patient's body for a multiple-field or moving-beam treatment is generally represented by isodose lines on body cross sections of the irradiated area. The cross sections of the body and organs must therefore be carefully drawn in a system of coordinates, if possible with the aid of computed tomography images. The calculation of isodose distribution with the aid of the computer, by the summation of the dose values of single fields, was described thoroughly by Poretti and Ionesco-Farca [1].

The distribution of the radiation dose inside the patient's body depends significantly on two factors: on the *curvature of the body* as well as the *shape and composition of the organs.*

The differences of the distance in the tissues covered by the electrons due to the body curvature cause a change in the single-field lattice (usually determined photographically on rectangular phantoms), which must be taken into consideration prior to the computer summation. The second correction of the fixed-field values becomes necessary because of the difference in absorption of the electrons by *body organs* with other chemical composition as the calibration medium, water.

Because of the *curvature of the body,* the precise mathematical determination of the differences in penetration of the electrons into the tissue is a very extensive task. It is possible, however, to show by experiment that relatively simple rules suffice in practice, like the one for gamma rays involving "air-gap shift rules" for deformation of the isodose curves [1, 2].

---

* Sumarized version read by Prof. D. Harder
** With the technical assistance of P. Ott, J. Feuz, and Fl. Fug

The mathematical calculation of the *changes in the isodose distribution within and behind an organ*, such as the lungs or bones (density other than 1 g/cm³) is also very extensive, and accurate results can really only be achieved with time-consuming computer simulation with expensive Monte Carlo methods. By using semiempirical methods it is nevertheless possible to obtain sufficiently accurate results for the requirements of radiotherapy.

According to a suggestion of Pohlit [3] the dose D'(z) at a body depth of z cm, taking into account the absorption effect of the organs, could be calculated as follows:

$$D'(z) = \left[\frac{f + z - d \cdot (1 - C)}{f + z} \right]^2 \cdot D \left[z - d \cdot (1 - C) \right]$$

d, path length in the organ, cm;

c, a coefficient to be experimentally determined ("coefficient of equivalent thickness"),

f, FSD, cm.

The application of this mathematical correction is limited to relatively large organs, larger than the radiation field (for small "inhomogeneities" see [4]).

The formula for high energy electrons with energy between 15 and 35 MeV of our betatron has been tested experimentally several times, i.e., using cork as "inhomogeneity." The "lungs" were simulated in a relatively large water phantom, using a piece of cork. For the coefficient c, as per Laughlin et al. [5], c = 1.3 X density of the cork. The calculated and measured (with calibrated Baldwin-Farmer ionisation chambers and Ionex device, Nuclear Enterprises Ltd, Edingburgh, UK) depth dose values behind a cork square (lungs), for different water (soft tissue) layers between "inhomogeneity" and body surface (1-3 cm), are in good agreement.

Organ Correction by Irradiation of the Human Thorax

General Remarks

The above-mentioned formula can also be applied for irradiation of the human thorax. The normal human lung tissue contains much more air than does cork, however, and by irradiation of the thorax other soft tissue lying on the surface, as well as the lungs and *ribs* will be penetrated by electrons. By empirical methods, namely by iterative comparison of experimentally determined dose values or isodose with the computer-calculated values of various densities ς, the optimal average *"thorax-density"* ς* can be derived.

As an experimental substitute of human thorax an Alderson phantom was irradiated; in which normal "lungs", consisting of a microcellular rigid foam with Z identical to soft tissue and density ς = 0.32 ± 0.01 g/cm³, were used surrounded by human ribs. The isodose distribution was determined by the usual photographic procedure (Kodak M films, in a Kodak Versamat 4 automatic developing machine, temperature controlled within 0.1 °C) using a body cross section at the ninth vertebra thoracalis and the absorbed dose in several points within the phantom, with the aid of thermoluminescent dosimeters.

The results contain several errors. For instance, total film blackening can only be obtained by compressing of the phantom slices in order to remove practically all the air bubbles between emulsion and phantom material. This is not always possible in the case of plates consisting of relatively soft material.

X-Ray pictures of the phantom sections indicate, moreover, that the phantom material (isocyanate rubber) is not fully homogenous, certainly not as are Polystyrol or Perspex, a fact which makes the measurements more difficult, particularly when measuring the depth dose curves in lung and bonefree areas of the phantom (soft tissue). These difficulties cause small differences between film dosimetry in the Alderson phantom and those isodose calculated with the computer (see Fig. 1).

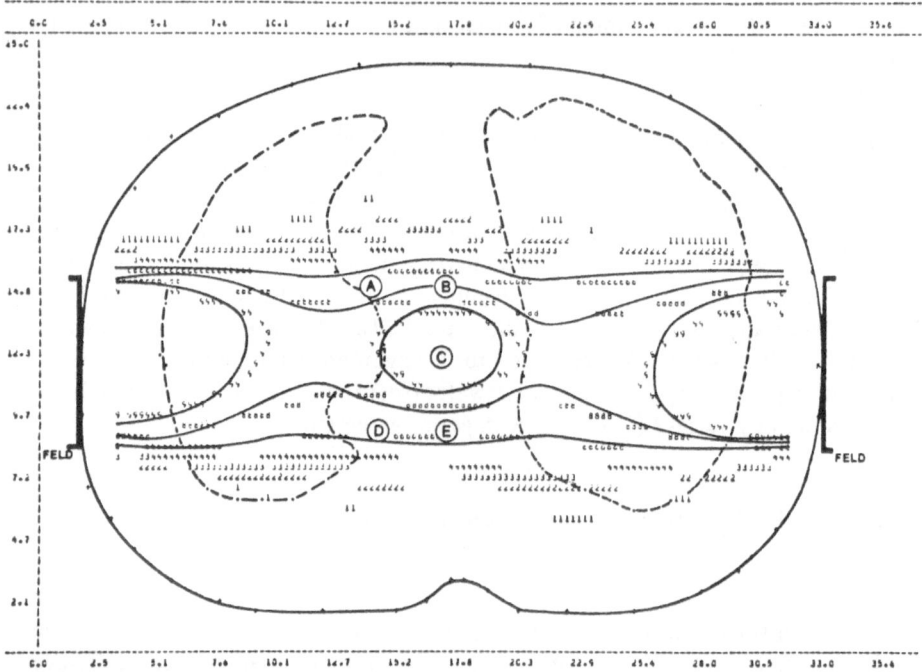

Fig. 1. Crossfield irradiation of the Alderson thorax with electrons (s. remark under "Experimental Results"). Energy of electrons, 30 MeV; field size, 7 x 10 cm. Calculation of the isodose with consideration of the correction due to the body curvature and of the thorax ($\varsigma^* = 0.4$ g/cm^3).

A-D indicate the points where the absorbed doses were measured with the aid of TL dosimeters.

Drawn lines, film isodoses (corrected ionometrically). Numerals, computer isodoses (8 = 80%), the dot here indicates where the isodose passes through. The position of the percentage figures obtained through interpolation seldom coincides with the mechanically determined position − 1/10 inch step in horizontal resp. 1/6 inch step in vertical direction of the computer's line printer)

Experimental Results

Considering the fact that the method is still in an experimental phase, the isodose plans shown here are examples only. The crossfield configuration of Figs. 1 and 2 for instance does not represent the usual clinical treatment method: it was chosen only so that the "worst-case" correction for lung could be demonstrated.

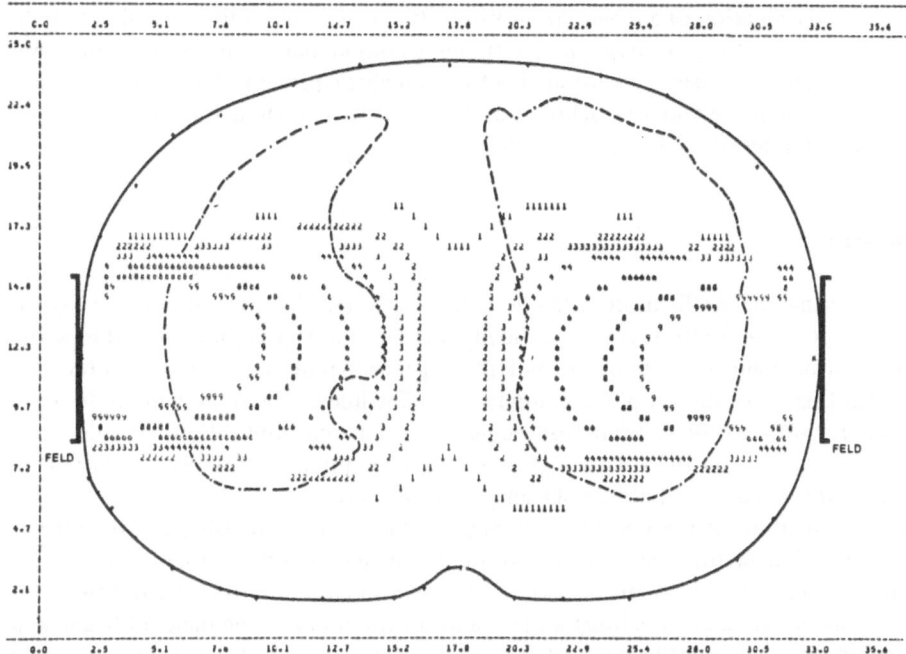

Fig. 2. Same crossfield irradiation as in Fig. 1 with consideration of the correction due to the body curvature but *without* correction for the internal parts of the thorax. Observe the difference from Fig. 1 of the dose distribution, for instance in the mediastinal area

Through comparison with the experimentally determined values, a value $\varsigma^* = 0.4$ g/cm^3 (instead of 0.32, a 25% higher value therefore) was used for the computer organ carrection.

With due consideration of the remarks made in the previous section and in the legend of Fig. 1, the agreement between the calculation and the experimentally determined values of Fig. 1 may be regarded as satisfactory. The comparison between Figs. 2 and 1 (i.e., computer calculations with due consideration of correction because of the patient's curvature of the body, but without correction for inhomogeneities in Fig. 2, and the calculation with additional consideration of the less strongly marked absorption of electrons in the lungs, in Fig. 1) clearly shows the clinical importance of the mentioned "thorax correction." As shown in Fig. 1, absorbed doses were measured in the points A, B, C, D, and E of the phantom by means of TL dosimeters (LiF rods, 1 mm ∅, 6 mm long).

The doses calculated from the single field values (from which the isodose points were derived by summation and interpolation, see above) agree within ± 5% with the ones that were measured.

It may therefore by assumed with a good measure of probability that the mentioned corrections for high energy electrons will result *in a clinically dependable dose distribution in the patient's body* in most cases.

The corrections that were made possible by computer methods demand the experimental determination of a *mean "thorax density"*, which may vary according to radiation apparatus, electron energy, bremsstrahlung contamination of the electron beam, etc., and also of course, as mentioned before, a realistic presentation of the cross sections (circumference and internal organs, if possible with a whole body computerized axial tomography of the body part to be irradiated).

Summary

The experimental verification of the dose distribution in the body of patients treated with high energy electrons is very difficult and inaccurate as, among other things, the body contours and the size of the patient's organs are seldom identical with those of the phantom. With the aid of the computer it is possible to elaborate quickly for each treatment, sufficiently accurate isodose plans for practical applications. The dose distribution depends on the following machine factors: number of fields or rotation angle, size of the radiation fields, FSD, and electron energy.

For the curvature of the body surface and the shape and composition of the organs, two correction factors have to be introduced in the computer program. First, one which considers the differences in the distance covered in the tissue by the electrons due to the curvature of the body surface and the obliquity of the incident beam; and a second one, necessary because of the varying absorption of the electrons by body organs of different composition.

For treatment in the thoracic region with high energy electrons, not only soft tissue and lungs are irradiated, but also ribs.

The correction factor has been determined experimentally in two ways: through simulation of the lungs with cork in a water phantom (water and lungs) and with an Alderson phantom (water, ribs, and lungs). Only the correction factor obtained with this latter experiment affords a clinically reliable computer picture of the dose distribution in the patient's body.

Acknowledgement. We are greatly indebted to the following persons and institutions for valuable advice and assistance during the extensive experimental work undertaken: Schweizerische Krebsliga (Swiss Cancer Foundation); Winterhalter Foundation, Zürich, Mendrisio (CH); Prof. Dr. med. A. Zuppinger, former director of our institute.

The co-workers of the Department for Medical Radiation Physics and of the Institute for Radiotherapy, University, Inselspital Bern (CH): Mrs. A. Pochon, Miss A. Giger, Dr. G. Garavaglia, Dr. R. Mini, as well as Drs. Guenda and Ezio Foglia for their assistance given on the basis of their doctors thesis.

References

1 Poretti GG, Ionesco-Farca F (1978) Use of the computer in high-energy electron therapy. Radiol Clin 47:139
2 National Bureau of Standards (USA) (1963) ICRU Report 10d. Clinical Dosimetry. Report Nr. 87. Superintendent of Documents, US. Government Printing Office, Washington D.C. USA
3 Pohlit W (1960) Dose distribution in inhomogeneous media by means of irradiation with fast electrons. Fortschr Röntgenstr 93:631-641
4 Abou Mandour M, Harder D (1978) Berechnung der Dosisverteilung schneller Elektronen in und hinter Gewebeinhomogenitäten beliebiger Breite. Strahlentherapie 154:546
5 Laughlin JS, Lundy A, Philips R, Chu F, Sattar A (1965) Electron beam treatment planning in inhomogeneous tissue. Radiology 85:524-531
6 Poretti GG (1975) Symposium on high energy electron therapy, Montreux 1964. Springer, Berlin Heidelberg New York

Electron Beam Quality Parameters and Absorbed Dose Distributions from Therapy Accelerators

A. Brahme and H. Svensson

Introduction

The quality of therapeutic electron beams is often specified by one single parameter, namely, the energy as determined from the practical range. However, the practical range is a parameter which is generally not directly related to the therapeutically useful part of the absorbed dose distribution. Therefore, further parameters are needed [1] to obtain a specification of the beam quality which is more relevant from a therapeutic point of view and, at the same time, simple.

Central-Axis Absorbed Dose Distribution

The distribution of absorbed dose in a medium irradiated by a therapeutic electron beam of given energy depends on a great number of secondary factors, such as field size, source-surface distance (SSD), composition of the medium, etc. To use a measuring situation as independent as possible of these parameters and to enable meaningful comparison to be made between electron beams from different accelerators, measurements should be made on the central axis in a water phantom at 100 cm SSD for large field sizes (i.e., > 10 x 10 cm at 10 MeV and > 15 x 15 cm at 40 MeV).

A number of features generally characterize the central-axis depth absorbed dose distribution of broad electron beams (see Fig. 1 of ref. [4][1]). With increasing distance from the phantom surface, the dose distribution is characterized by (a) the near-surface dose, (b) the dose build-up of secondary electrons, (c) the dose build-up of primary electrons due to increased inclination of the electron tracks [3], (d) the dose maximum where the electrons approach full diffusion [3], (e) the descending section with increased loss of the primary electrons, (f) the steep section of linear dose decrease due to energy and range straggling of the primary electrons, (g) the tail of electrons having suffered few interactions in the first parts of their tracks, and, finally, (h) the photon background contributed by photons generated both in the phantom and in the accelerator. Several of these features will appear in the following definition and comparisons of parameters of therapeutic and physical interest.

Comparison of Electron Beam Quality Parameters from Therapy Accelerators

Dose Gradient (G). The most important feature of the electron depth-dose curve is the very steep decrease in dose beyond the therapeutic range. This rapid dose fall-off is a

---

1 Most of the figures which have been left out, due to editorial problems, can be found in a less updated version in references [4] and [8]

great advantage particularly when there are organs at risk behind the tumor that must be protected from unnecessary irradiation. A useful measure of the steepness of the absorbed dose distributions is the dose gradient along the central axis of the beam. To obtain a measure of the dose gradient which is, as far as possible, independent of the energy and the photon contamination of the beam, a normalized dose gradient has been defined [4, 8] by multiplying the absolute dose gradient by the practical range and dividing it by the maximum dose due to electrons

$$G = \frac{R_p}{D_m - D_x} \left| \frac{dD}{dz_{max}} \right|.$$

From a known depth absorbed dose distribution, the dose gradient can be obtained simply by extrapolating the steepest section of the dose distribution, which is used to obtain the practical range, until it intersects the photon background (D_x) and the level of maximum absorbed dose (D_m). The dose gradient is now obtained by dividing the practical range by the distance between these two intersections. This measure of the steepness of the absorbed dose distribution is a dimensionless number which varies slowly with electron energy since the increase in practical range with energy almost balances the decrease in absolute slope of the absorbed distribution.

In Fig. 1 the dose gradient is plotted as a function of the most probable electron energy for seven different accelerators and according to theory (Monte Carlo calculations [4]).

It is observed that the dose gradient for most accelerators differs significantly from the theoretical values, the situation being worst for the standing wave linear accelerator which is known to have a wide energy spread. The best agreement with theory is obtained for the scanning beam linear accelerator and the dual scattering foil [5] flattened microtron.

Practical Range (R$_p$). The practical or projected range (R_p) is the range concept in most frequent use for correlation with the electron energy and, as it has long been in use, it needs no detailed discussion here. The energy calculated from the practical range is most closely related to the most probable electron energy at the phantom surface.

Half-Value Depth (R$_{50}$). The depth at which the absorbed dose has decreased to 50% of its maximum value has been suggested by the HPA [6] as a measure of the energy of high-energy electron beams. However, on theoretical grounds, the half-value depth should be expected to be related more closely to the mean energy than to the most probable energy of the incident electrons. Poor agreement is also found between energy determinations based on the practical range (R_p) and the half-value depth (R_{50}) [7]. Brahme and Svensson [4] showed that the variation of the half-value depth with the mean electron energy at the phantom surface for different betatron and microtron beams is linear up to 30 MeV. These accelerator types was chosen because the mean energy can be calculated easily from the almost monoenergetic energy (E_a), of their accelerated electrons. Even when the mean energy differs by several MeV from the most probable energy, a very close linear relation between the mean energy at the surface and the half-value depth is obtained below 30 MeV:

$$R_{50} = c \, \overline{E}_0$$

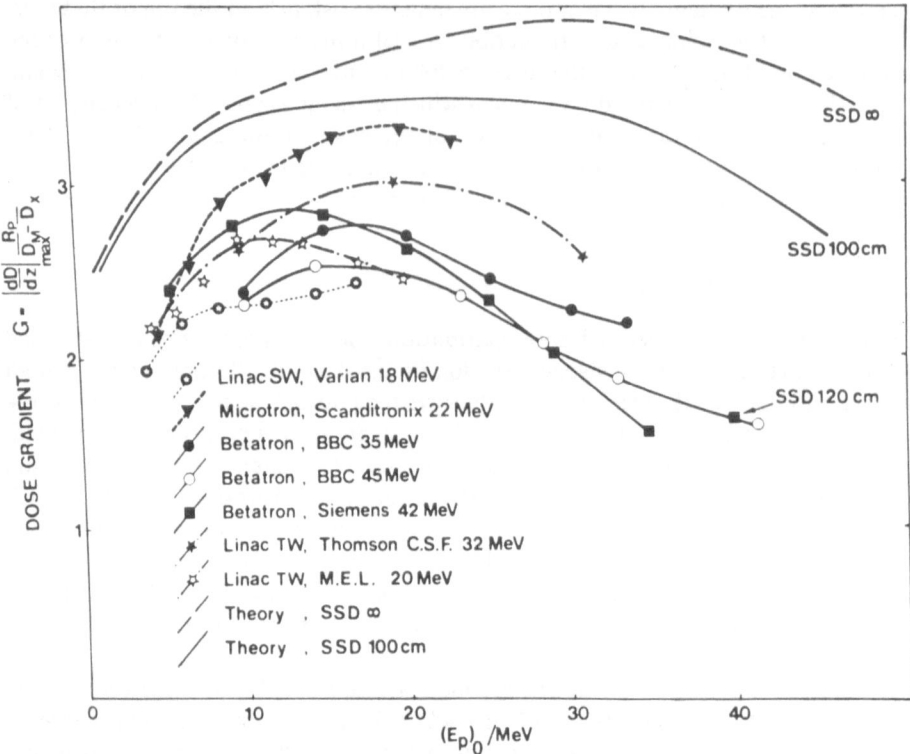

Fig. 1. Dependence of the dose gradient, G, on the most probable electron energy at the phantom surface for monoenergetic beams [2] and beams from therapy accelerators. The symbols correspond to the following accelerators: star in circle, Varian 18-MeV Standing wave linac; solid triangle, Scanditronix 22 MeV microtron; solid circles, BBC 35-MeV betatron; open circles, BBC 45-MeV betatron; solid squares, Siemens 42-MeV betatron; solid stars, Thomson C S F 32-MeV travelling wave linac; open stars, M E L 20-MeV travelling wave linac. All curves are for SSD 100 cm and large fields unless otherwise stated

where $c = 0.43$ cm MeV^{-1} at infinite source suface distance. This relation can therefore be used in a first approximation to determine the mean energy at the surface also in linear accelerator beams where the energy distribution sometimes is quite broad but still of the skew-straggeling type obtained in thick foils. The mean energy determined in this way is of dosimetric importance as it is the basis for the selection of stopping power ratios when the composition of the dosimeter probe differs from that of the medium.

Therapeutic Range (R_{85}). The therapeutic range is that depth interval of the absorbed dose distribution which ideally should coincide with the target volume. For a given electron energy, the width of this interval depends on the desired uniformity of the absorbed dose distribution in the target volume. It is well known that differences as small as 5%-10% in the absorbed dose in some cases can cause a considerable change in the difference between the fraction of patients cured and that subsequently suffering complications, a fact which normally makes it desirable to strive for a very small dose variation inside the target volume.

The uniformity of the absorbed dose distribution perpendicular to the beam can sometimes be improved by increasing the field size somewhat. Similarly, the distribution of absorbed dose parallel to the beam can often be improved by increasing the electron energy. However, both these methods result in a larger irradiated volume, a result which is generally not acceptable. If the same variation in absorbed dose is accepted, perpendicular and parallel to the beam, the variation of absorbed dose in the target volume should be less than 10%, since most accelerators are specifed with a homogeneity of ± 5%. Such as small variation with depth is generally unrealistic since a compromise has to be found between good uniformity and a small irradiated volume. The therapeutic range has therefore been defined as the depth interval within which the absorbed dose exceeds 85% of its maximum value. For most high-energy electron beams, the therapeutic range extends from the surface to the 85% depth doese, since the surface dose is often higher than 85% of the maximum dose. In Fig. 7 of ref. [8] three quite different depth dose curves are shown all with the same therapeutic range of 7 cm. It is seen that the therapeutic range of the single scattering foil betatron is not increased by increasing the energy since, simultaneously, a thicker foil is needed which leaves the therapeutic range approximately unchanged. However, the dose behind the therapeutic range will increase considerably.

In Fig. 4 of ref. [4] the therapeutic range is plotted as a function of the most probable electron energy for the same beams as before. It is observed from this diagram that the therapeutic range above approximately 20 MeV can not be improved by simply increasing the accelerator energy [8]. Again, the best agreement with theory is obtained for the scanning beam linar accelerator (full stars) and the microtron (full triangles).

Surface Dose (D_S). The entrance dose in a volume irradiated by a therapeutic electron beam is generally of great interest since it is of importance for the degree of skin sparing obtained. The depth of the radiation-sensitive layers below the epidermis is generally of the order of 0.5 mm, a depth which is also accessible for accurate absorbed dose measurements with many detectors. The surface or skin dose, D_S, is thus defined here as the ratio of the absorbed dose at a depth of 0.5 mm to the maximum absorbed dose.

The surface dose is often about 90% or less because of the buildup of the fluence of primary and secondary electrons and in most therapeutic beams also because of the contamination by oblique electrons and photons.

In Fig. 7 of ref. [4] the surface dose is plotted as a function of the most probable electron energy. The skin sparing is lost in most therapeutic beams even though it should be present, at least below approximately 15 MeV. In one betatron the skin sparing is better than theoretically expected due to an excessive photon background (cf. Fig. 8 of Ref. [4]).

Photon Background (D_X). The photon background (D_X) is defined as an extrapolation of the tail of the absorbed dose distribution back to the practical range. The photons are either present as a contamination of the incident electron beam or generated in the irradiated medium itself.

In Fig. 8 of ref. [4] theoretical results are shown for monoenergetic and monodirectional beams, and they therefore only include the photons generated in the medium. The experimental results from therapy beams contain a considerable fraction of contaminating photons, since these beams are generally scattered by metal foils and other constructional details which produce bremsstrahlung. The scanning beam linear accelerator and the dual scattering foil flattened microtron show only a negligible increase above the theoretical photon background generated in the phantom.

Influence of Accelerator Design on the Distribution of Absorbed Dose

Electron beams from therapy accelerators differ in three principal ways from the plane parallel and monoenergetic beam which would give ideal dose distributions for most therapeutic purposes. In clinical electron beams the electrons are first of all distributed over a wide energy range; secondly, their direction of motion is spread over a large angular interval; and finally, the uniformity and beam size is often far from the infinite uniform beam. The origin of these differences is, of course, partly to be found in the properties of the crude electron beam delivered by the accelerator, but all the materials in and along the beam-like vacuum windows, scattering foils, transmission monitors, and air are normally of even greater importance. To conclude the influence of these three factors on the shape of the depth-dose curve will be briefly reviewed.

Field Size and Uniformity. The influence of field size on the shape of the depth-dose curve has been studied during many years. The loss in therapeutic range and skin sparing is well known when the diameter of the field is decreased [10]. One factor which has been less extensively studied is seen in Figure 7 of this reference [10]. At 10 MeV in water, broad beam conditions are obtained at a diameter of 7 cm; still there is a great difference in the build-up region between a 7-cm and a "infinite" uncollimated beam. This difference is due to the electrons entering the collimator from the source side but later scattered out through the collimator edge [9]. These electrons are of great importance for small field sizes and near the borders of large fields.

Furthermore, a larger number of electrons are scattered out at a collimator edge for low-Z (Z = atomic number) materials than for high-Z material. This is caused by the longer lateral movement of the electrons in low density materials and consequent larger area along the edge of the collimator that can contribute with edge-scattered electrons. The best results is therefore obtained with the highest possible density at the beam edge e.g. an aluminum collimator laminated with 1 mm tungsten [9].

The width of the border region z_1 [3] which contributes 99% of the edge-scattered electrons is in fairly good agreement proportional to the measured amount of electrons scattered out through the collimator edge. It is also of interest to observe that beyond dose maximum no influence of the electrons scattered through the edge can be observed [10]. This can be shown to be caused by the low mean energy of these electrons, being only 40% of the mean electron energy of incident beam [10].

Consequently, the edge-scattered electrons must be regarded as a beam contamination since they have no effect at all from dose maximum down to the therapeutic range, even though they may somewhat increase the uniformity at the surface of a badly flattened beam.

Angular Distribution. The influence of the angular spread of an almost monoenergetic electron beam on the shape of the depth-dose curve is clearly demonstrated in Fig. 2.

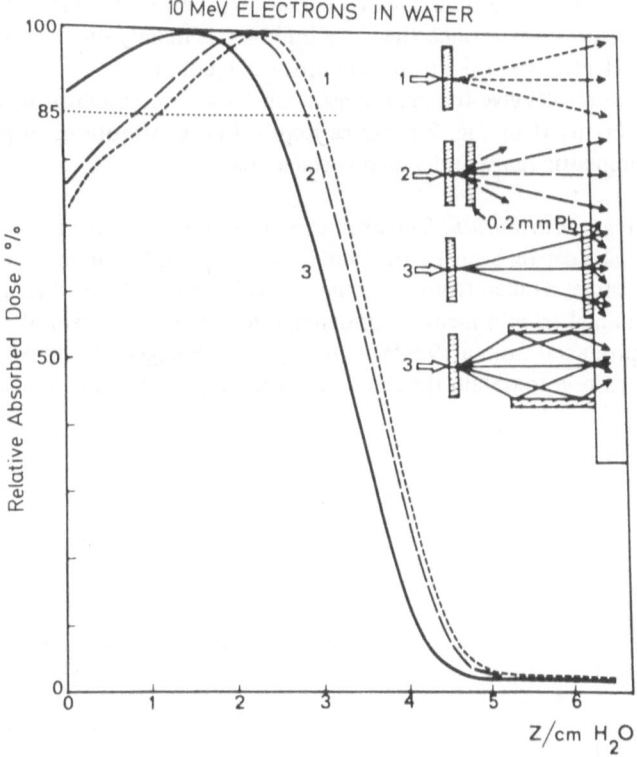

Fig. 2. Comparison of depth-dose curves from electron beams of different angular spread. For beam 3 the root mean square angular spread of the electrons is about 20°, but for beam 2 it is only a few degrees; other parameters like energy spread and beam size are identical. The angular spread of the electrons reaching the surface is the main reason why scatter collimators of the tube type produce degraded depth-dose curves

Curve 1 is the central axis depth-dose distribution of a 10 MeV clinical electron beam. In curve 2 the scattering foil thickness has been increased by 0.2 mm lead which has decreased the practical range due to energy loss and decreased the therapeutic range even more due to energy straggling in the foil. The angular distributions of the electrons in these two beams are almost identical as the extra lead foil was placed together with the normal scattering foil. In curve 3 the extra scattering foil has been moved and placed at the phantom surface. The energy distributions of curves 2 and 3 are thus almost identical. However, the angular distributions of the electrons that reach the phantom are completely different. The electrons in curves 1 and 2 are almost parallel as they represent the central 10 cm of a divergent beam from a point source at 100-cm distance. In curve 3 the angular spread of the electrons reaching the surface of the phantom is rather wide ($\Theta_{rms} \approx 20°$) due to the foil at the surface. This has resulted in a decreased practical range, since the mean angle of incidence of the electrons is increased ($R_p' \approx R_p \cos \Theta_{rms}$). The loss in therapeutic range is even larger because the electrons reach a state of full diffusion much more quickly when they already have, at the surface, a wide angular spread [9]. This also explains the smaller dose build-up near the surface since the dose build-up is mainly due to the increase in obliquity of the electrons as they penetrate into the medium.

These results give the main reason why scatter collimators of the tube type illustrated in the insert of Fig. 2 produce depth dose distributions of poor quality with a small therapeutic range and a high surface dose.

Energy Distribution. The influence of the energy distribution of an almost monodirectional electron beam on the shape of the depth dose curve is shown in Fig. 3. The dashed curve is taken from a normal clinical beam with small energy spread, $\Gamma = 0.2$ MeV, obtained from a nearly monoenergetic accelerator beam of 14.4 MeV. The full curve is produced from a 19.9 MeV, nearly monoenergetic beam which has been decelerated by carbon to obtain the same practical range as for the clinical beam.

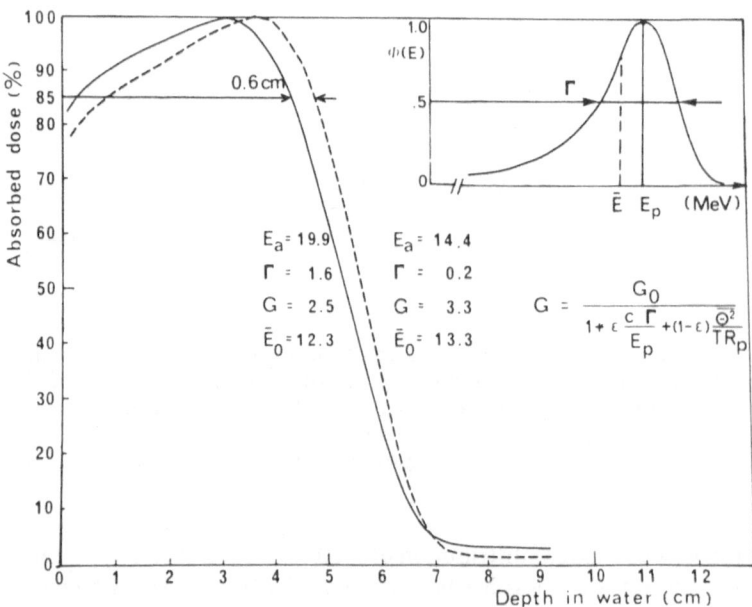

Fig. 3. Comparison of depth dose curves from electron beams of different energy spread but almost identical, most probable energies. The inserted equation relates the dose gradient of a degraded beam to that of a monodirectional and monoenergetic beam (G_0). The constants, ϵ, c, and T depend on the phantom material: T being the mass scattering power, c the ratio of most probable energy loss to energy straggling, and ϵ the relative importance of energy and angular spread. For low atomic number phantom materials, c is close to 5 and ϵ is about 0.45

The wider energy distribution obtained in this way has a clear influence on the shape of the depth curve (compare the insert with $\Gamma = 1.6$ MeV). Fundamentally, it is the slope of the dose fall-off that has decreased, as one would expect, since the energy straggling in the beam is transformed to a range straggling in the phantom. As a matter of fact, it is possible to derive a simple expression which relates the dose gradient (G) to the energy spread (Γ) and the mean square angular spread ($\overline{\Theta^2}$) for broad electron beams [8]. This relation is also included in the figure.

The energy spread of more than 1 MeV was introduced here somewhat artificially by a thick carbon decelerator. However, an energy spread of this order of magnitude is not uncommon in clinical beams because the energy spread of the accelerator alone may reach these values for many linear accelerators, particularly of the standing wave type, and so may the energy spread produced in scattering foils and other materials in the beam particularly if the beam is broad and the energy is high [4].

Conclusion

By decreasing the energy and angular spread the quality of the electron beams from most existing treatment units can be improved considerably with regard to factors such as dose-gradient, uniformity, build-up characteristics, and photon contamination. These factors are becoming of increasing clinical importance, because improved diagnostic techniques, with transmission and emission tomographs and ultra sound, are becoming available, and not least because increased radiobiologic knowledge about dose fractionation and the very steep dose response relations of the tumor and normal tissues are steadily obtained.

References

1 de Almeida, CE, Almond PR (1974) Radiol 111:439
2 Berger MI, Seltzer SM (1969) Paper presented at the 12th Int. Cong. Radiol. Tokyo, Japan
3 Brahme A (1975) Thesis, University of Stockholm
4 Brahme A, Svensson G (1976) Med Phys 3:95
5 Brahme A (1977) Paper presented at the 14th Int. Cong. Radiol. Rio de Janeiro, Brasil
6 HPA (1971) (Hospital Physicist Association) Rep Ser No 4
7 de Almeida CE, Almond PR (1973) Phys Med Biol 18:737
8 Brahme A, Svensson H (1979) Acta Radiol. Ther. Phys. Biol. Acta radiol 18:244
9 Brahme A (1978) Paper presented at the Annual Meeting of the Swedish Association of Physicists in Medicine, Örebro
10 Lax I, Brahme A (1980) (to be published in Acta Radiol)

Surface Dose in Electron Beams and Association with High Energy X-Ray Beams

J.C. Rosenwald*

Electron treatments give skin reactions far more severe than high energy X-ray treatments at equal doses. Sometimes these reactions may become even too severe for the treatment to be continued at the dose initially planned. This paper describes how the surface dose is modified according to the different treatment parameters and what can be done to keep it as low as possible.

Theory

Electronic build up does not exist in electron beams as it does in high energy photon beams: the incident electrons begin loosing their energy just as they penetrate the medium. However, one can observe a slight dose build up in the first millimeters because of two major reasons [1]:

1) The incident electrons create a number of secondary electrons more numerous at depth than at the surface.
2) The incident electrons reach the skin perpendicular to the surface and they are scattered in the first millimeters of tissue. Therefore, because of the increasing scattering angle, the energy loss is larger for deeper layers resulting in an increase of dose.

The second phenomenon is predominant; it is more pronounced for smaller electron energies. At greater depth, the dose decreases because of the loss of electrons scattered out of the beam or because they have been stopped. The dose is maximal at a depth which can be theoretically calculated. It has been found that at 100 cm source-skin distance (SSD), this D_{max} depth is between 2 and 5 cm for energies between 8 and 40 MeV [2]. Such calculations also give the surface dose (as defined at 0.5 mm depth) which is found to be between 68% and 92% of D_{max} at the same SSD and for the same energy interval [2, 3].

Actually, the measured surface dose is generally higher [3, 4]. This can be explained partly by the width of the spectrum of the accelerated electrons and partly by the same contamination by low energy electrons and photons scattered from the scattering foil, monitor, and beam limiting device. For these reasons, low surface doses are observed in quasi-monoenergetic beams (such as in microtron), in scanned electron beams without scattering foil (such as in Sagittaire), and for thin collimating systems.

---

* We are grateful to Miss C. Mazeau for having performed most of the experimental work

Methods of Measurement

An appropriate dosimeter has to be chosen for measurements at small depths: this dosimeter must be thin enough, and it must have a response independent of the large spectral variations found in the first millimeters. Flat air ionization chambers with thin walls and, preferably, extrapolation chambers can be considered as thin enough, as well as photographic emulsions, and thin thermoluminescent disks [5-8].

As far as the energy response is concerned, liquid ionization chambers and ferrous sulfate dosimeters are more suitable [9, 10].

Table 1. A comparison of the thicknesses of dosimeters [5-10]

Dosimeter	Wall thickness (mm)	Sensitive volume thickness (mm)
Flat air ionization chamber	0.03	0.5
Extrapolation chamber	0.07	–
Liquid ionization chamber	0.3	0.3
Ferrous sulfate	0.1	0.8
Single coated photographic film	–	0.03
Double coated photographic film	0.2	0.03 + 0.03
Teflon-lithium borate disks	–	0.13

Experimental Results

We have performed a number of comparative measurements with flat ionization chambers and photographic films in the beam of a BBC Asklepitron 35 betatron. The measurements were carried out in a polystyrene phantom for different electron energies and a number of cone sizes. A negative bias voltage (250 V) was applied to the flat chamber after it had been demonstrated that reversing the polarity did not affect the reading by more than 1%. Photographic measurements were made with double coated Gevaert-Structurix D2 films, tightly pressed, without cover, in the phantom and exposed parallel or perpendicular to the beam axis (see Table 2).

There is a good agreement between flat chamber and perpendicular film results; but parallel films, more difficult to handle properly, show a lower response. This underestimation is more pronounced at higher energy. Such a result, previously found by several authors, had been tentatively explained by Markus and Paul [11].

The surface doses, as measured with the first two methods, are in good agreement with the values reported earlier for the same betatron fitted with standard cones; according to the theory, the surface dose increases with energy. It can also be noticed how important the contamination is for small energies and small cone sizes (almost 10% difference at 14 MeV with and without the $\emptyset$ 4 cone).

We have already discussed the possibility of reducing the skin dose (adjunction of thin collimator, removal of scattering foil, reduction of the spectrum width); however, the lower limit, as predicted by the theory, is still quite high, especially at high energy.

Table 2. Comparison involving photographic films and a flat ionization chamber

$(Ep)_0$ (MeV)	Cone (cm)	D_{max} depth (cm)	Surface dose (% of D_{max})		
			Flat chamber (%)	Film perpendicular (%)	Film parallel (%)
14	∅ 4	0.7	92.3	91.7	86.0
	12 x 14	1.8	88.5	90.0	80.0
	none	1.8	83.0	85.3	86.0
28	∅ 4	1.8	93.7	93.2	79.8
30	12 x 14	1.8	93.3	92.5	80.5
	none	2.0	90.3	89.0	82.4

Surface Dose in Photon Beams

In high energy photon beams, the surface dose is very low. As an example, the surface dose measured in the 30 MV X-ray beam of our betatron has been found to be 6% of D_{max} for a 5 x 5 cm field, at 100 cm SSD, without blocking tray. However, it shows large variations depending upon SSD, collimator setting, flattening filter, and existence of blocking tray. Figure 1 illustrates the shape of the depth dose curve in the first centimeters for two extreme situations. Figure 2 shows the surface dose variation as a function of field size for different flattening filters. It can be seen that this dose is always much less in our RX beam than in our electron beam.

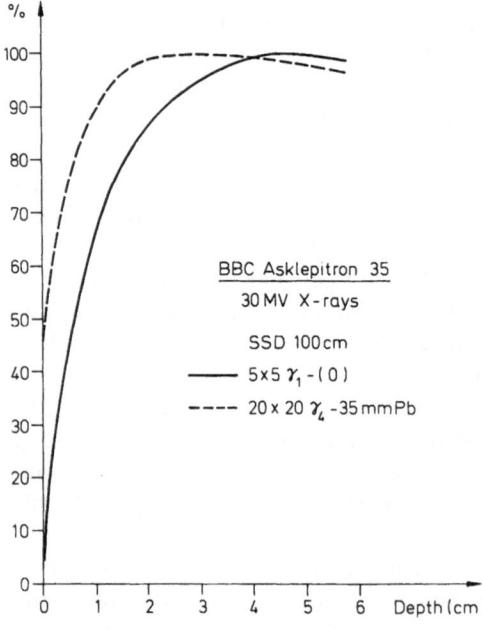

BBC Asklepitron 35

30 MV X-rays

SSD 100 cm

——— 5x5 γ_1 - (0)

---- 20 x 20 γ_4 - 35 mm Pb

Fig. 1. Depth dose curves at small depth for: small field size and no flattening filter ($\gamma 1$); large field size and thicker flattening filter ($\gamma 4$)

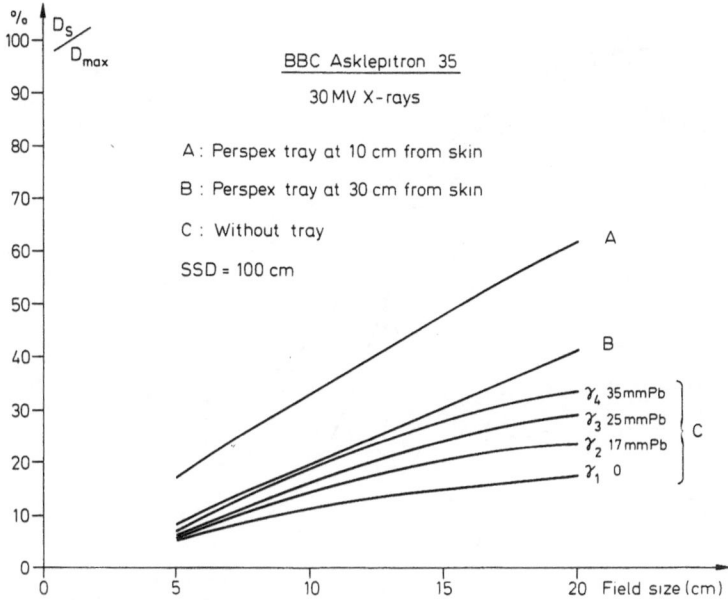

Fig. 2. Surface dose as a function of field size, with and without blocking tray and for different flattening filters

Photon-Electron Combination

As a result of this observation, one way to reduce skin dose in an electron treatment is to combine it with a high energy photon treatment through the same portal [12]. Obviously, such an association leads to a dose enhancement at great depth which depends on the relative contribution of the X-ray beam. This contribution has, therefore, to be adjusted, as well as the electron energy. An example of the resulting depth-dose curve is given in Fig. 3 for 30 MeV electrons and various weighting factors. In most cases the added photon beam is chosen to contribute to D_{max} as only one-third of the electron beam. This provides a surface dose at least 20% lower and a dose beyond the target volume 10% to 20% of D_{max} higher. The advantages of a photon-electron combination are more pronounced for high energy electron beams: for such beams the surface dose is higher and the depth dose curves show less distortion as the photon beam is added.

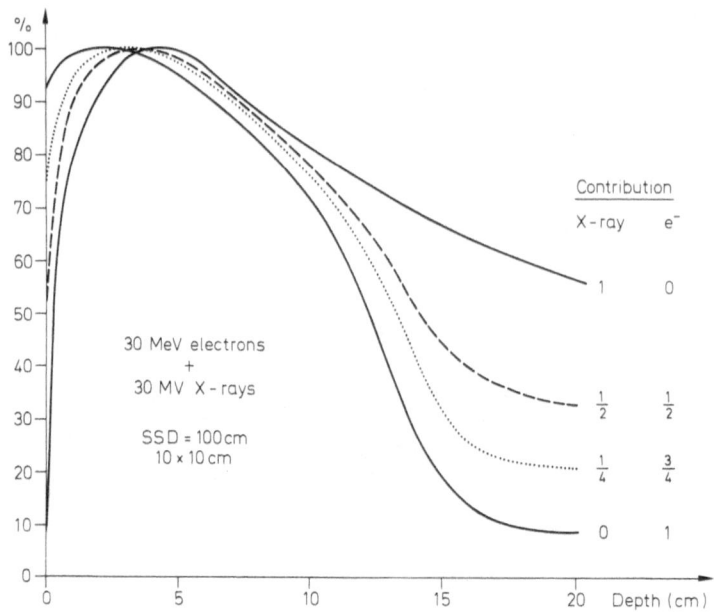

Fig. 3. Depth dose curves for different electron-photon combinations

References

1   Blanc D, Dutreix A, Mathieu J (1974) In: Physique de la Radiothérapie. Presses Universitaires de France, P 141
2   Berger MJ, Seltzer SM (1969) Calculation of energy and charge deposition and of electron flux in a water medium bombarded with 20-MeV electrons. Ann NY Acad Sci 161:8-23
3   Brahme A, Svensson H (1976) Specification of electron beam quality from the central axis depth absorbed-dose distribution. Med Phys 3:95-102
4   Brahme A, Hultén G, Svensson H (1975) Electron depth absorbed dose distribution for a 10 MeV clinical microton. Phys Med Biol 20:39-46
5   Morris WT, Owen B (1975) An ionization chamber for therapy-level dosimetry of electron beams. Phys Med Biol 20:718-727
6   Järvinen H (1977) Extrapolation chamber measurement of electron depth absorbed dose distribution. Acta Radiol (Ther) 16:129-135
7   Dutreix J, Dutreix A (1969) Film dosimetry of high-energy electrons. Ann NY Acad Sci 161:33-43
8   Chavaudra J, Marinello G, Brulé AM et al. (1976) Utilisation pratique du borate de lithium en dosimétrie par thermoluminescence. J Radiol Electrol Med Nucl 57:435-445
9   Hultén G, Svensson H (1975) Electron depth absorbed doses for small phantom depths. Acta Radiol (Ther) 14:537-544
10 Svensson H, Hettinger G (1967) Measurement of doses from high-energy electron beams at small phantom depths. Acta Radiol (Ther) 6:289-293
11 Markus B, Paul W (1953) Über Dosismessung und Dosisverteilung in elektronenbestrahlten Körpern. Strahlentherapie 92:599-611
12 Gharbi H (1970) Mémoire CES Radiologie. Paris

Electronic Wedge Filter for the Asklepitron 45

R. Hünig, A. v. Arx, and A. Scholz

A wedge filter placed in the useful beam of a megavolt therapy unit produces a tilting of the dose distribution to the beam axis. Appropriate superimposition of several fields can produce a uniform dose in the volume to be treated. Wedge filters are useful in the treatment of tumors in the region of the head and the neck. Their use is, however, not restricted to these areas. Wedge filters are used to influence the isodoses of moving beams as well.

The deformation of isodoses by means of wedge filters is required in the treatment with high energy (25-45 MeV) electrons as well as with photons. The former is, amongst others, to be preferred in all cases where an extensive sparing of structures behind a lesion is necessary. The problem with high energy electron irradiation has so far been solved by using physical wedge filters made of wax or other materials inserted in the applicator next to the skin. The disadvantages of this procedure are:

1) Dependence on energy, field size, each case requiring the preparation and testing of a separate wedge filter for the desired isodose deformation.
2) The wedge filter can be inserted in the applicator only when all parameters, excepting the focus skin distance (FSD), are set by means of optical centering devices.
3) The wedge filter placed in near contact with the skin removes the build-up effect.
4) An air gap between wedge and skin reduces the effect of the wedge filter.
5) With the interposition of a wedge filter in the electron beam an inhomogeneity of the energy spectrum is produced, caused by the energy losses, dependent on thickness. This causes an energy spread of the radiation, increasing with the size of the wedge.

All the above-mentioned disadvantages of mechanical wedge filters can, however, be overcome by using electronic wedge filters. The objective is the same, i.e., to get the required dose gradient over the entire beam width. The apparatus required for such filters is a part of the normal equipment of the BBC 45 MeV betatron. The principle is shown in Fig. 1. The magnetic field of a pair of coils acts on the pulse of monoenergetic electrons exiting from the accelerating tube and, at the same time, influences the distribution of the electrons in the plane of the tube. This distribution, after the electrons have passed the dual scattering system, is checked by differential dose measurements on both sides of the beam and is kept constant by means of an electronic regulator; it is normally kept symmetric with respect to the central axis of the collimating system. To obtain wedge filter type isodoses it is sufficient to modify the electronic regulator by means of the preselection controls placed on the main control desk, resulting in a defined controllable asymmetry in the loading of the two dose-measuring electrodes. A displacement of the 80%-isodose of 45° can be obtained in this manner (Fig. 2). Small fields limit the maximum obtainable wedge filter effect, as do low energies (Fig. 3).

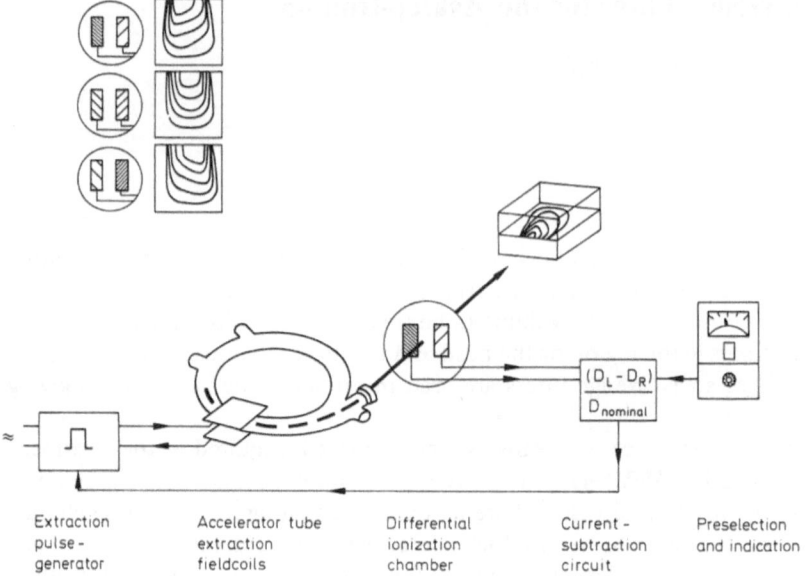

Fig. 1. Principles of the Asklepitron 45 wedge filter

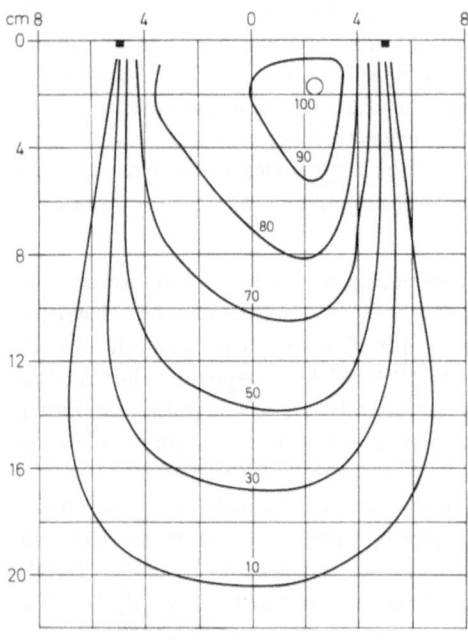

Fig. 2. Electronic wedge filter — Asklepitron 45. Electron beam isodose curves, radiographic measurement. Energy, 45 MeV; field size, 10 x 8 cm; FSD, 110 cm

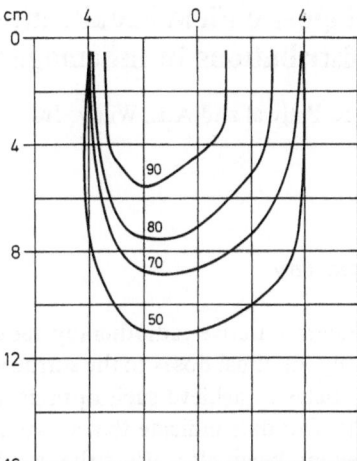

Fig. 3. Electronic wedge filter —Asklepitron 45. Electron beam isodose curves, radiographic measurement. Energy, 30 MeV; field size, 8K x 10 cm; FDS, 110 cm; dyssymmetry, max.

To set the right priorities in the review of the advantages of the electronic wedge filter, the order of the already mentioned disadvantages of the wax wedge filter will be as follows:

1) The energy spectrum is absolutely homogeneous at right angles to the central beam. The wedge filter effect will, as in the case of the effect of wedge filters for photon radiation, be reached only by means of different dose rates perpendicular to the central beam.
2) Under usual treatment conditions an air gap between localizer edge and skin does not influence the effect of the wedge filter.
3) The build-up effect remains.
4) The setting of the electron beam wedge filter is obtained electronically. It also can be reproduced via automatic control, at a later date.
5) The carrying out of an electronic wedge filter irradiation requires only a slight amount of extra time on the part of the radiographer. The setting of the unit can be made by means of the localizer and with help of the light indicator.
6) The relations between the various parameters which determine the dose distribution and the degree of the asymmetry that can be chosen are, once they have been determined and measured, constant and can consequently be tabulated. The setting of wedge filter fields therefore, requires no more time than that needed for open fields.

The advantages of high energy electron radiation have been repeatedly pointed out by most radiotherapists. If the decision is made to adopt the indication on electron therapy the described electronic wedge filter can contribute considerably to improve the treatment techniques.

Magnetic Field Modification of Electron Beam Dose Distributions in Inhomogeneous Media*

B.R. Paliwal and A.L. Wiley, Jr.

Summary

Modern curative radiotherapy requires higher doses to the tumor volume and, necessarily, minimal doses to the surrounding normal tissues. Attempts to use heavy charged particles to achieve such optimization are currently under investigation in many centers. Our data indicate that a static, superimposed magnetic field on a clinical electron therapy beam also offers the capability of some "tailoring" of isodose distributions. Furthermore, a *variable*, superimposed magnetic field minimizes those tissue generated dose heterogeneities which are inherent with all charged particle beams. We suggest that magnetically modified, clinically available electron beams also offer a practical and less expensive means of achieving tailored, heterogeneity-corrected, isodose distributions.

* The original paper has been published under the following title: Paliwal BR, Wiley AL, Wessels BW, Choi MC (1978): Magnetic field modification of electron-beam dose distributions in inhomogeneous media. Med Phys 5, 404-408

Conclusions of the Physical Section

J.C. Rosenwald

The papers presented in the physical section have demonstrated the important improvements obtained during the past few years: they concern the electron beam characteristics as well as the dose calculation procedures.

The most significant technical improvement lies in the possibility of getting very clean electron beams such as those from microtrons or from "scanning beam" linear accelerators. This improvement has been evaluated quantitatively (A. Brahme) consisting mainly in a diminution of skin dose and a better protection of tissue around the tumor. Other methods for adjusting the dose distribution have been discussed: combination of electron beams and high energy X-ray beams (J.C. Rosenwald), installation of isodose shaping magnets (B. Paliwal, G. Weishjer); but they have a smaller impact on the general improvement.

As far as the dose calculation procedures are concerned, the weak point remains the methods of accounting properly for body inhomogeneities. The CT scanners now make possible an accurate reconstruction of the body structures (shape, position, and density). It is therefore fundamental to set up good quality correction algorithms. The work presented here on this subject (D. Harder, G. Poretti) is very encouraging.

Now, what is still to be done? It seems to me that two kind of risks are associated with electron treatments:

1) *The risk of an error in dosimetry.* This risk, very high at the beginning of electron therapy, has been considerably reduced through a better knowledge of the beam characteristics and a standardization of the procedures commonly used. However, it is mandatory to be extremely cautious in a number of circumstances; this has been explained in detail by several participants (A. Dutreix, J.L. Minchole).

2) *The risk of an error in interpretation.* This risk exists when a treatment technique in current use in one hospital is adopted by another hospital or when clinical results from different treatment centers are reviewed. It consists mainly in a misinterpretation of energy and dose values selected for reporting.

It has been clearly demonstrated (A. Brahme) that "energy" itself can not be considered as representative of the treatment depth. The same energy for the same patient treated with the same dose can give a good dose distribution with one accelerator and a significant tumor underdosage with another one. It is therefore much better to specify a treatment depth rather than an energy. Usually, the energy is chosen in order to reach a given dose level (generally 80% or 90% of the peak dose) at a given depth. A common language would be established if we could agree on the isodose value to be selected in order to choose the proper energy: the 85% isodose does not differ much from what us used in most hospitals (80%-90% range); it seems a reasonable compromise between a good dose homogeneity within the tumor and a rapid fall-off beyond the target volume. The choice of the 85% depth, named as "therapeutic range", can therefore be strongly recommended to specify the electron beam penetration; hopefully, in the near future, it will not be said any longer "this patient has been treated with 15 MeV electrons" but "this patient has been treated with 4.2 cm electrons."

Dose specification is also somewhat confusing: we have heard during this symposium a nice mixture of "rads", "grays", and "centigrays". I think that this is all right; using simultaneously both systems is a good way of getting progressively used to the new one. However, another confusion is much more serious, i.e., in most circumstances, the radiotherapists have reported the patient doses, without stating clearly what they meant. For some of them, it was the maximum target dose (100%); for others, the minimum target dose (80% or 90%). Here again a common language is necessary. The best seems to be to stick to the ICRU recommendations given in a recent report (Report 29: "dose specification for reporting external beam therapy with photons an electrons"). These recommendations consist, for electrons, in specifying the dose on the 100% isodose (peak absorbed dose). They are, therefore, very simple, and I wish that everyone would accept them.

Clinical Section

Clinical Radiobiology

A. Zuppinger

In the second and third decades of this century the French school at the Curie Foundation, Paris, under the direction of Regaud, and also Coutard, who was a chief of the X-ray department, brought radiotherapy to a new basis. Clinical radiobiology was the main factor which enabled the change in the radiologic treatment of cancer. The X-ray doses were always measured, but the reaction of the normal tissue and the shrinking of the tumor were decisive for the final dose.

Today one applies more or less systematically 6000-7000 rads in 5-7 weeks, which simplifies the procedure and facilitates some statistical studies, but it is very questionable whether this procedure is optimal. When we began 20 years ago with electron therapy, it was recommended to increase the dose by 10% in comparison to that of photons. We recognized very early that the reaction of the normal tissues and the shrinking of the tumors were less pronounced in comparison to that with photons, and we saw that with higher energies a supplementary dose had to be given. These observations were interpreted as relative biological effectiveness (RBE) which changed with the energy with values of 0.8-0.7 and even less in the high energy range of 30 to 35 MeV. We were forced to make the dosimetry in the sense of a *biophysical method*. The publication of Report 21 in 1972 by the Bureau of standards was a great help, since it allowed the calculation of the rad on an exact basis. The dependence upon the voltage disappeared practically completely, but we had still to increase the dose by about 10% in comparison to that of photons.

In my opinion, one should not renounce biology. The schematic dosimetry does not consider the individual differences in the sensitivity of the normal tissues, which fluctuate ± 15% about a mean value. The responses of the tumors differ even more. There exists a parallelism between the radiosensitivity of the normal tissues and the tumors which arise from them. There are only a few exceptions to this rule. It can best be recognized in tumors of the mucous membranes. Patients with early and strong reactions show, generally, an early shrinking of the tumor (Fig. 1). On the other hand, one sees patients with lack of or only a slight reaction from the normal dose. If one applies the normal dose in the patients with early reactions, one runs a high risk of radiation damage, while in the others the tumors do not disappear, or an early recurrence occurs. In the latter case, the dose may be raised without increased risk of damage. If the tumor did not shrink at the expected rate, we generally interrupted the treatment at the onset of the reaction and applied an additional dose of between 3000 and 4000 rads at the end of the reaction. In several cases we added radium needling. This procedure has already been developed with conventional X-rays (1951) [2] and was later on called split therapy. Split therapy is recommended only in cases where a treatment in one series seems not to be successful, also in elderly patients or in those with a poor general condition where a severe reaction represents an enhanced danger.

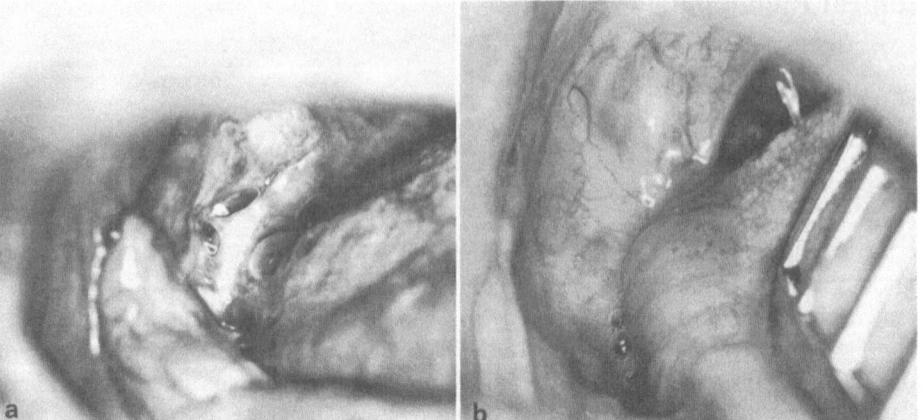

Fig. 1 a, b. Extensive retromandibular tumor T_3. Irridation with 4390 rads (1500 rets). The tumor disappeared completely. The patient was free of disease until death 7 years later because of prostatic cancer. a Strong reaction after 3760 rads, 3 days before the end of the treatment. b 3 years later. Normal appearance of the retrobuccal region

Clinical radiobiology shows with high speed electron treatment a surprising effect: good cure rats in some tumors which were considered to be radioresistant. It is possible to control the tumor in a high percentage of patients — often in combination with surgery — with parotis tumors and soft tissue sarcomas, as we will show later on. Especially astonishing are the results in *synoviomas,* which are considered to be highly radioresistant, so that in the great majority of cases amputation can be avoided. Also in *Hürthle cell carcinomas,* a tumor which has been encountered much more frequently in the last decade and which was considered to be highly radioresistant, local control is possible in the majority of patients, while with photons no cure occurred. We also have some chordomas and adamantinomas which respond well, so that in those patients in whom surgery is not possible, a trial with electrons is justified.

Gastrointestinal tumors could, up to the present time, be influenced only in an unsatisfactory way, but there is a better response to electrons, as will be shown later on.

In patients with fixed lymph nodes some surprising cures have been seen. One may object that this is also possible in some cases with photons. We made a comparison of photons and electrons which showed better results with electrons. The effect is so good, that we renounced entirely the prophylactic neck dissection in patients with oral cavity tumors.

It is surprising that the same kind of tumors and tumor localizations which respond better to electrons than photons can also be better controlled by neutrons, as Dr. Catteral [1] states. These facts can be explained neither by the absorbed dose nor by the effect upon the anoxic cells. It may be that a different distribution of the ionozations is responsible for these effects.

But there is an important clinical difference: by treating these tumors with electrons, the amount of radiation damage is lower than with neutrons. Electron treatment has the additional advantage that one can treat tumors which arise in *radiosensitive tissues.* This is possible in tumors of the outer genital region, mainly in penis carcinoma, and in metastases in the inguinal region, even if they are ulcerated. A corresponding situation exists in the field of the treatment of tumor recurrences after irradiation alone and after combined operation and irradiation, which will be discussed separately.

Electron therapy also has its limits, as with every therapeutic procedure in the field of malignant tumors until the drugs are found which cure anarchistic growth.

In tumors of the gastrointestinal tract electron therapy can — with the exception of very radiosensitive tumors — only be applied as a *preoperative treatment*, because of the high sensitivity of the mucous membranes and the peritoneum.

In preoperative treatment we have to distinguish clearly between two different procedures.

1) Preoperative treatment of the inoperable tumor extension with the goal of rendering the tumor operable; high doses are necessary and the interval between irradiation and operation must be long, between 4 and 8 weeks, because of the latency of the radiation effect.
2) Preoperative irradiation in operable cases with the goal of reducing the danger of propagation during the operation; the necessary dose is low, the interval between irradiation and operation only 2-3 days.

In the first mentioned treatment the complication is increased and there seems to be no effect on reducing the propagation rate. With the second method neither the operative procedure nor the postoperative phase is complicated. However, we recommend leaving the sutures intact 2-3 days longer than usual.

As long as we do not distinguish between these two procedures, we will never be able to prove that preoperative radiologic treatment is a useful procedure and should be applied on a broader scale.

In this brief survey of clinical radiobiology I have shown you my experiences with this new tool in the battle against cancer. Many treatment results are certain (some others are probable, and still others are possible) to contribute to progress in cancer therapy. We will still encounter many problems, and I hope you understand why I proposed in my opening speech that we should try, or even that we are obliged, to find a way to contribute to an early answer of the open questions so that we can determine whether the good results mentioned are only due to good luck or whether they correspond to real progress.

Summary

Since 1972 the dosimetry in high speech electrons is more exact, but an RBE of about 10% still has to be considered. A biophysical method is recommended in order to take the individual difference in the sensitivity of the normal tissues and the tumor into consideration. The indications for split therapy are discussed. Some tumors considered to be more or less radioresistant can be controlled, some of them in a high percentage. Also some special localizations of gastrointestinal tumors can be controlled, but here only preoperative irradiation is recommended. Also some tumors in radiosensitive tissues can be cured.

References

1 Catterall M, Beweley DK, Sutherland J (1977) Second report of a randomized clinical trial of fast neutrons. Br Med J 1:1642
2 Zuppinger A (1951) Biological problems in X-ray therapy of intrinsic and extrinsic tumors of the larynx. J Faculty Radiol 3:10-23

Indications for Electron Beam Therapy

J.P. Bataini

At the Curie Institute, a 35 MeV Brown Bovery betatron has been in clinical use as a source of fast electrons since June 1962. Because of the physical characteristics of the absorption of fast electrons in tissues, and as no difference has been found in the RBE of electrons and high energy photons, treatment with electrons was restricted to superficially or half deeply seated tumors.

Because of insufficient skin sparing, mixed beams are sometimes used with different ratios of ^{60}CO (5.5 or 30 MV photons) and (15, 20, or 25 MeV) electrons. The general indications for electron beam and mixed beam therapy at the Curie Institute are:

Cancer of head and neck
Cancer of breast
Cancer of penis, anal canal, Bowen disease of vulva
Certain cases of sarcoma of bone and soft tissue
Certain cases of brain tumors
Spinal cord irradiation in children
Extensive or recurrent skin carcinoma of face and scalp
Melanomas
Retinoblastoma and tumors of the orbit
Auditory canal, middle ear, and glomic tumors
Parotid carcinoma and preauricular nodes
Cancer of nasal fossa
Cancer of ethmoid and maxillary sinuses with anterior extension
Cancer of oral cavity and anterior tonsillar region
Metastatic cervical nodes

Head and Neck Cancer

The location of many neoplastic conditions of the head and neck region favors the employment of electron beam therapy at its best energy levels, i.e., up to 20 or 25 MeV in most cases, thus affording by a single portal technique a good and sometimes an almost ideal dose distribution. This, in a department dealing with a large number of head and neck cases, is definitely a very important advantage over other treatments that can be given by wedge filtered photon techniques at 2-6 MV energy levels or in occasional cases by extensive implants or radioactive applicators.

Based on a personal experience of 16 years, the electron beam has indeed proved to be very useful in the following malignant conditions of the head and neck region:

The management of tumors of the eye and orbit, cancer of the oral cavity and anterior tonsillar region, cancer of the paranasal sinuses, and finally metastatic cervical nodes will be dealt with in some of the following papers.

Other tumors in the head and neck area such as: laryngeal and pharyngeal cancers are still best managed with photon techniques. Cancer of the mobile tongue is more adequately treated by radioactive implants, very often combined with external photon beam therapy.

Radiation Management of Breast Cancer

The electron beam offers many advantages in the radiation management of breast cancer. Dose distribution at high level is practically homogeneous from skin to the selected depth in the tissues with a sharp and rapid fall-off and consequently an almost complete protection of the underlying deeper structures, provided the appropriate energy of the beam is chosen. The skin dose, which is related to the energy of the beam, is of the order of 85% of the applied dose, so that build up is not usually necessary unless the skin itself is definitely involved.

Electron beam could be used postoperatively after radical or simple mastectomy and in the treatment of chest wall recurrences. Low energy levels are necessary in these cases as also in the irradiation management of advanced inoperable flat breasts. Electron beam has also proved useful for booster doses in case of exclusive irradiation after the basic ^{60}CO loco-regional irradiation. This has proved very useful especially in peripherally located primary tumors (parasternal, high upper quadrants, extreme tail of breast, inframmary sulcus) where additional irradiation through reduced tangential breast fields with or without compression could not be achieved satisfactorily. It has also proved useful in low-seated axillary nodes.

The cosmetic results in patients receiving these supplementary electron beam treatments, however, seem inferior to those obtained in patients treated exclusively by ^{60}CO irradiation.

Conclusions

It has been said that electron beam therapy has no advantage over the treatments than can be given by supervoltage photon techniques and, in certain other cases, by interstitial implants. However, although dose distributions are in many cases comparable, the employment of fast electrons, up to 20-25 MeV, in a department dealing with a large number of superficially or half deeply seated tumors seems to be a definite advantage because of the much easier planning and setting up of patients. In certain conditions, electron beam therapy is indispensable in the management of the irradiation treatment. Energy levels higher than 25 MeV are infrequently used, preference being then given to multiportal photon techniques, which give better dose distributions.

The indications for electron beam must rely solely on dose distribution characteristics rather than on a very hypothetical particular biological effect.

The Electron Beam Therapy for Malignant Tumors: Indications and Limitations

E. Scherer and M. Bamberg

The purpose of this paper is to demonstrate the indications for electron beam therapy and to give a review of its role in modern cancer treatment. The basic concepts of clinical electron beam physics were described by Becker and Schubert [2], followed by the Symposium for High Energy Electrons, Montreux [18] and the summarizing reports of Zuppinger [17, 19], and Weitzel [15]. Two years ago Tapley gave a detailed synopsis of clinical applications of the electron beam in her monograph [13]. The present report is based on 12 years experience using a betatron with energies ranging from 5 to 43 MeV electrons and a 5.7 MeV linear accelerator, and also 20 years experience with a 18 MeV machine.

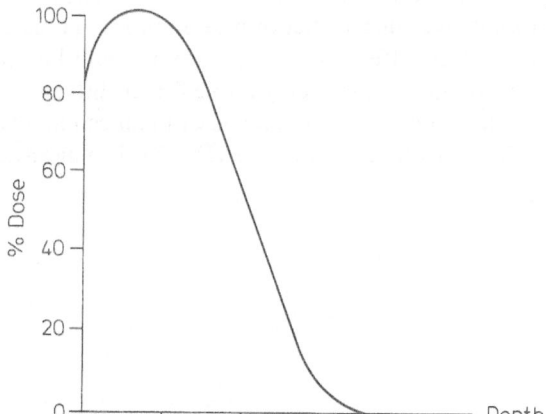

Fig. 1. Tumor and target volume, optimization of depth dose curve using fast electrons

For many tumors, the depth dose and dose distribution of electron beams are favorable, because the dose decreases rapidly beyond the depth determined by the energy of the beam. Tissue behind the tumor is better protected and the volume dose is lower (Fig. 1). In contrast to photon beams the dose build-up is smaller and the skin sparing effects are decreased (Fig. 2). The surface receives as much as 95% of the maximum dose at 18 MeV. Owing to varying density and different structures of the irradiated tissue, dose inhomogeneities of electron beams can be measured. Especially in compact bone, depending on its thickness, the dose is much more attenuated, so that a corrected dose has to be computed. According to Wideröe [16] the supposed tumor electivity can be increased by electron beams. However, in accordance with laboratory studies and clinical observations of some authors [4] these effects appear to be caused by the volume factor. In comparison to photon beams severe skin reactions are said

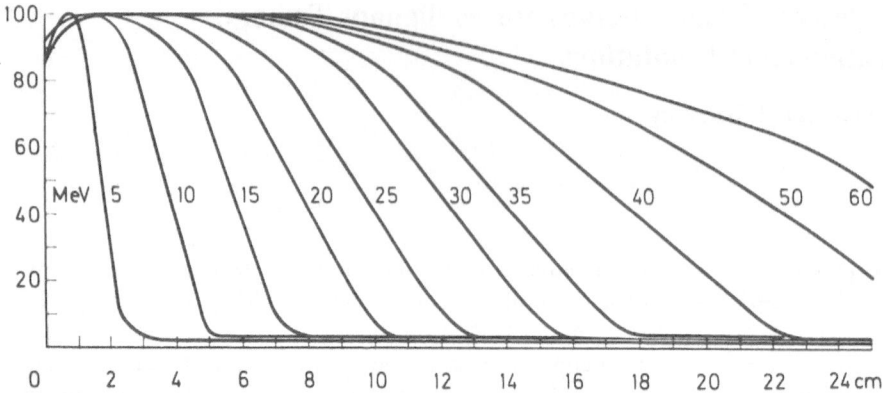

Fig. 2. Depth doses of fast electrons from 5 to 60 MeV

to heal better, and supposedly the same biological effect can be attained with fewer fractions per week, in small fields 3 x 3 Gy/week or 2 x 4 Gy/week, especially for palliative indications. The relative biological effectiveness ranges between 0.7 and 0.8 in the same level such as that of photon beams, at least to the 40% isodose [3]. However, a protracting effect of electron beams could be proven, which in contrast to the external irradiation plays only a role for the interstitial [198]gold application.

In certain tumors, such as rectal or brain carcinoma, electron beams can be favorably combined with photon beams (Fig. 3). The disadvantage of electron beams regarding

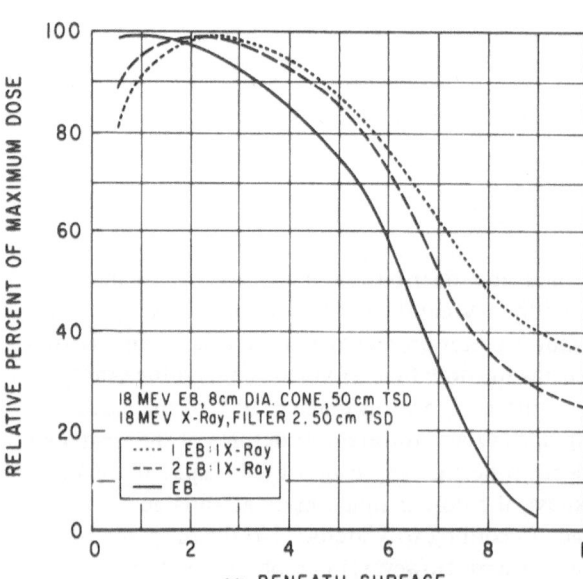

Fig. 3. Percent depth doses with combined electron and photon beams (Tapley [12])

dose build-up and bone inhomogeneities can be avoided by rotation techniques, especially by the telecentric irradiation with a small angle (Fig. 4). This technique, which we developed in our department [9], has proved to be useful in treatment of tumors of the anterior mediastinum (Fig. 5), small bladder lesions and in tumors of the para-

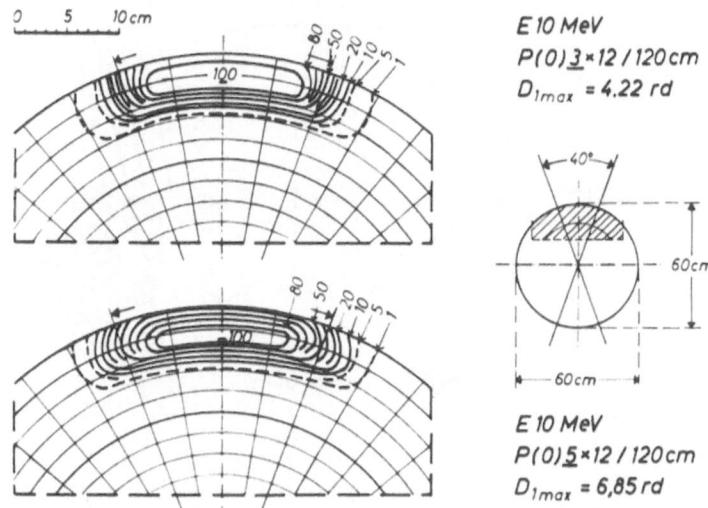

Fig. 4. Principle and depth doses of the telecentric arc therapy with small angle (30°-40°) (Rassow [9])

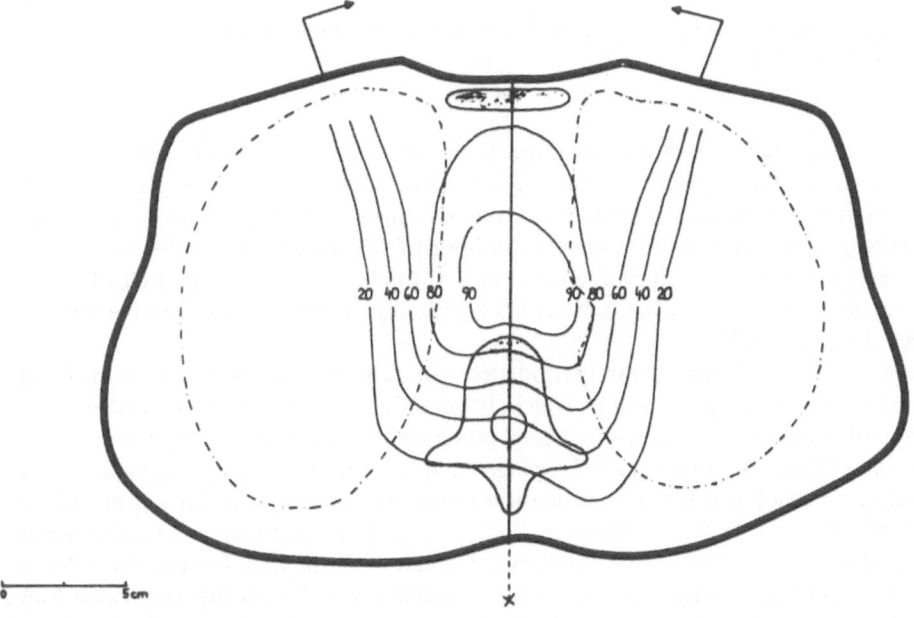

Fig. 5. Example of the telecentric arc therapy for tumors of the mediastinum (40°, 25 MeV, 12 x 3 cm portal) (Gürtler et al. [6])

metrium (Figs. 6, 7). In case of extensive lymphangiosis of breast cancer patients after radical mastectomy the rotation technique represents an interesting variation in the treatment of chest wall (Figs. 8, 9).

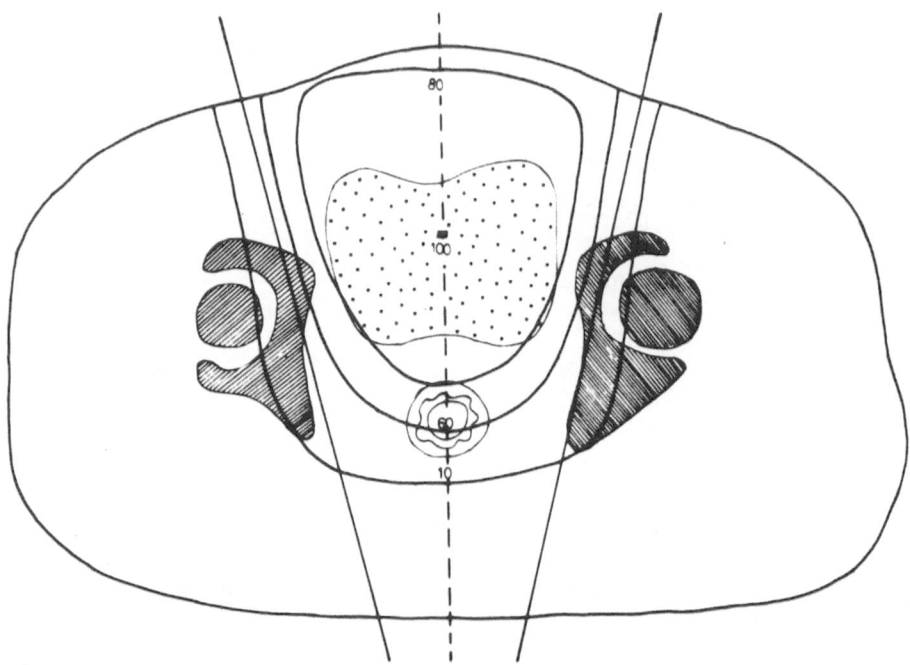

Fig. 6. Example of the telecentric arc therapy for a bladder carcinoma (30°, 30 MeV, 3 x 12 cm portal)

Good therapeutic effects with electron beam therapy can be achieved in squamous cell carcinoma of the head and neck. The tumor dose can be increased effectively by an electron beam boost in carcinoma of the oro-, meso-, and hypopharynx, especially in supraglottic lesions and tumors of the floor of the mouth and of the base of the tongue [11]. We have achieved good results with electron therapy in T_3-T_4 lesions, which are treated in five fractions of 1.5 Gy weekly in combination with bleomycin, 3 x 7.5 mg per week.

Carcinoma of the larynx is rarely irradiated with fast electrons, but some authors [8] could obtain similar good results, which, however, are not superior to telecobalt therapy. Tobin et al. [14] also reported good cure rates of head and neck tumors.

We can confirm the findings of Zuppinger that in treatment of parotid gland tumors electron irradiation is superior to photon beams. In orbital cancers it appears to be questionable wether fast electrons are indicated [18]. In these cases telecesium therapy should be given because of the better protection of the lens. In order to obtain a homogeneous dose distribution telecesium is applied in tumors of the nasal cavity and of the large spindle cell carcinoma of the skin with the exception of extensive cancer of the lip. In recurrent cervical lymph node metastases optimal dose distribution can be

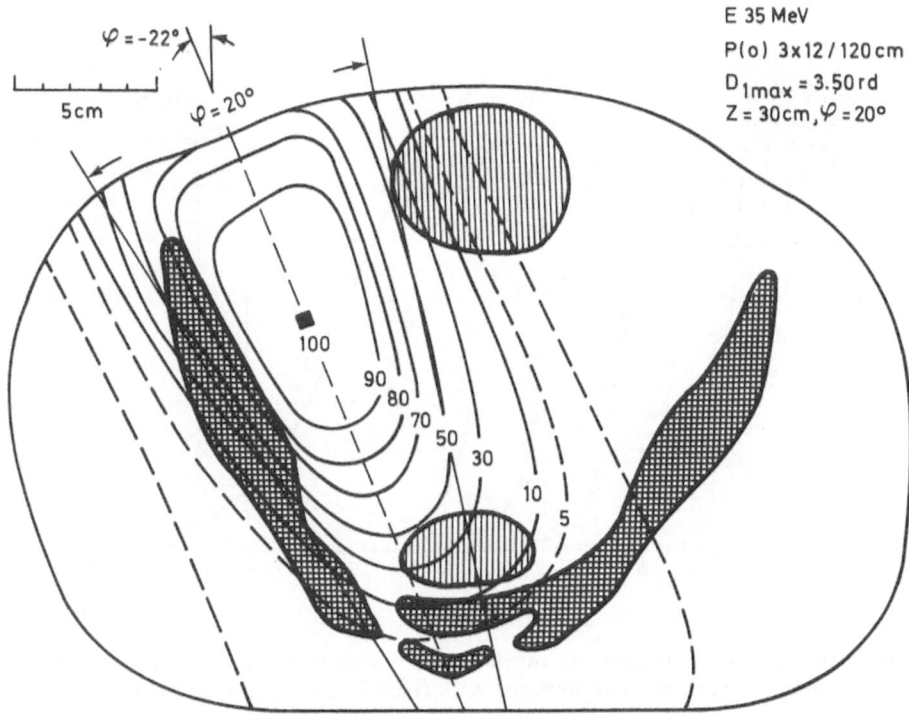

Fig. 7. Example of the telecentric arc therapy for a parametric residual mass of cervix carcinoma

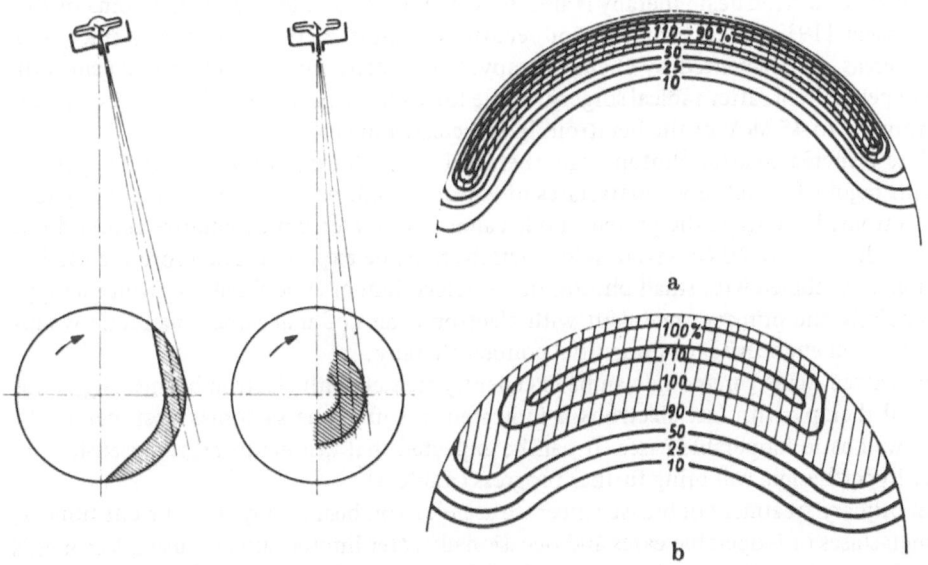

Fig. 8. Excentric arc therapy for lymphangiosis carcinomatosa of the chest wall (*a* 5 MeV; *b* 25 MeV) (Weitzel [15])

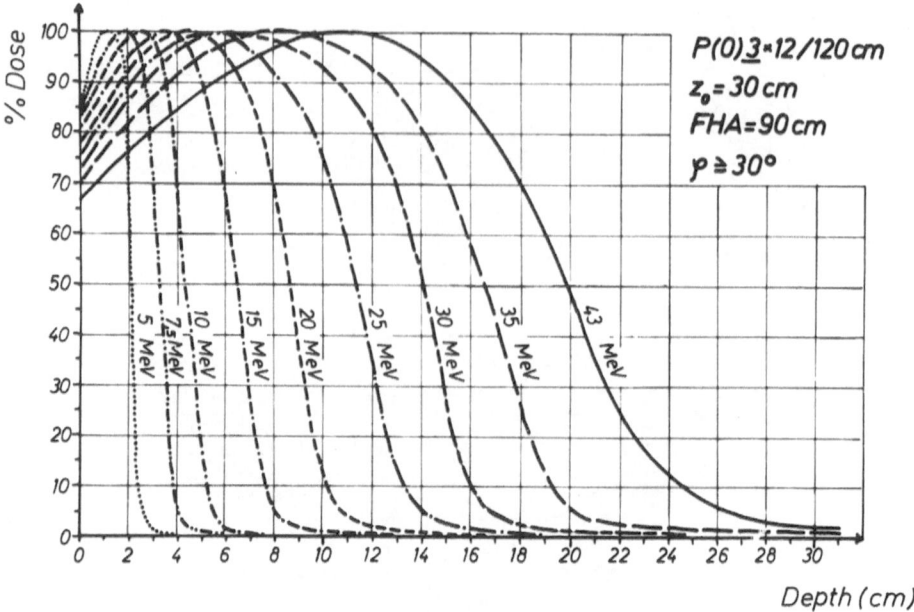

Fig. 9. Depth doses of telecentric arc therapy show an increased build-up effect and a steeper slope of depth curves compared with fixed beam therapy (Rassow 1974)

achieved [10]. The advantage of electron beams is evident in treatment of thyroid cancer, since the cervical spinal cord can be spared.

Palliative electron beam therapy is also indicated in cases of localized carcinoma of the stomach [19], smaller primaries or recurrent tumors of the liver, biliary tract, and pancreas, supplemented by chemotherapy. We irradiate the recurrent rectal cancer of the pelvic floor, after radical surgery, in the knee-elbow position with energies ranging from 30 to 45 MeV of the betatron. In some cases such as prostatic cancer the tumor dose is increased after photon beam therapy by an electron boost through the perineum. Inguinal lymph node metastases of vulvar carcinoma should be irradiated by fast electrons. In contrast the primary node cannot be treated with a curative tumor dose, as in doses above 40 Gy severe skin reactions must be expected. Therefore the irradiation is combined with small photon fields, telecesium or superficial roentgen therapy. Similarly the primary treatment with electron beams of anal carcinoma requires supplemental endocavital or interstitial contact therapy.

Advanced penis carcinoma can be excellently treated with electron beams. High dose local therapy with fast electrons is indicated in soft tissue sarcomas postoperatively as well as in inoperable cases. It can be expected that fast neutrons, radiosensitizers, or hyperthermia will bring further progress (Table 3).

In primary treatment of breast cancer we use electron beams only in recurrent tumors, metastases or inoperable cases and occasionally after limited surgery, using Veronesi's technique (so-called quadrantectomy), although fast electrons offer a wide range of individual applicability [13].

Indisputable indications for electron irradiation are invasive carcinomas of the skin, malignant melanoma and its lymph node metastases, and mycosis fungoides. In spite of the progress achieved by chemotherapy total body irradiation with low energy fast electrons represents the choice of treatment for the different stages of the mycosis fungoides [7].

Table 1. Indications for electron therapy alone

Tumor region	Telecesium therapy possible
Skin ⎱ Lip ⎰ greater lesions	−
Melanoma	−
Mycosis fungoides	−
Soft tissue sarcoma	−
Lymph nodes (inguinal, axillary)	(+)
Vulva	(+)
Parotid	+
Mamma IV	+
Nose, ear	++
Mamma I (Op. with Veronesi's technique)	++
Penis	++

Table 2. Indications for use of electrons as boost after primary photon beam

Tumor region	Tumor region
Cerebral tumors	Cervical lymph nodes
Oropharynx, tonsil	Thyroid
Hypopharynx	Esophagus (upper third)
Epipharynx	Mamma I (RTX alone)
Base of the tongue	Recidive of rectum carcinoma
Mouth floor	Prostate

We are convinced that in the future a cancer center should have a particle accelerator with an energy up to 25 MeV. Absolute and relative indications for the electron therapy are summarized in Table 1 and 2. Especially, in palliative treatment of locally advanced tumors electron beam therapy grants the best dose distribution and therefore a good protection by a small integral dose. An intraoperative irradiation technique has been described by some Japanese authors [1]. At presently we are comparing the effects after radiotherapy with fast electrons and neutrons, especially in cancer of the head and neck. Current and planned studies in our center are shown in Table 3.

Table 3. Trials started or in preparation

---

1. Comparison between electron therapy and fast neutrons for soft tissue sarcomas, carcinoma of the salivary glands, recurrences of tumors of the bladder, rectum, and prostate.
2. Combined treatment modality with electrons and misonidazole (see the References in [5a]) (malignant glioma, carcinoma of the esophagus, carcinoma of the vulva, soft tissue carcoma, malignant melanoma)
3. Electron beam therapy in combination with hyperthermia (T_{3-4} carcinoma of the head and neck inclusive recurrences, malignant melanoma)

References

1 Abe M, Takahashi M, Yabumoto E, Onoyama Y, Torizuka K, Tobe T, Mon K (1975) Techniques, indications and results of intraoperative radiotherapy of advanced cancers. Radiology 116:693
2 Becker J, Schubert G (1961) Die Supervolttherapie. Thieme, Stuttgart
3 Botstein Ch (1976) Clinical application if high-energy (18-35 MeV) electrons. In: Tapley du VN (ed) Clinical applications of the electron beam. J. Wiley, New York
4 Degner W, Fürst G (1977) Gedanken zum gegenwärtigen Stand der Strahlentherapie mit hochenergetischen Elektronen. Radiobiol Radiother 18:459
5 Dimopoulos J, Kärcher KH, Peloschek P, Epp N (1976) Die Stellung der Megavolttherapie im Behandlungskonzept des Prostatakarzinoms. Z Urol Nephrol 69: 489
5a Dische S, Saunders MI, Flockhart, IR, Lee ME, Anderson P (1979) Misonidazole − a drug for trial in radiotherapy and oncology. Int J Radiat Oncol Biol Phys 5: 851
6 Gürtler KF, Darai S, Schnabel K, Kuttig H (1977) Elektronentiefentherapie im Thoraxbereich. V., Dosimetrische Untersuchungen mit telezentrischer Kleinwinkelbestrahlung. Strahlentherapie 153:143
7 Hoppe RT, Fuks Z, Bagshaw MA (1977) The rationale for curative radiotherapy in mycosis fungoides. Int J Radiat Oncol Biol Phys 2:843
8 Piquet JJ, Desaulty A, Pillaert JM, Lefevre JL, Barrois J, Decroix G (1976) Le traitement de cancers glottiques. J Belge Radiol 59:319
9 Rassow J (1976) Beitrag zur Elektronentiefentherapie mittels Pendelbestrahlung. IV. Mitt.: Über eine neuartige, für primär unaufgestreute Elektronen spezifische telezentrische Kleinwinkelpendeltechnik. Strahlentherapie 140:156
10 Scherer E, Rassow J (1971) Methodische Grundlagen der perkutanen Strahlenbehandlung von Lymphknotenmetastasen des Halses. Strahlentherapie 141:523
11 Spanos WJ jr, Schukovsky LJ, Fletcher GH (1976) Time, dose and tumor volume relationships in irradiation of squamous cell carcinomas of the base of the tongue. Cancer 37:2591
12 Tapley du VN (ed) (1976) Clinical applications of the electron beam. J. Wiley, New York
13 Tapley du VN, Montague ED (1976) Elective irradiation with the electron beam after mastectomy for breast cancer. Am J Roentgenol 126:127
14 Tobin DA, Hall W, Scott RM (1976) Electron beam therapy in head and neck tumors. Am J Roentgenol 126:1251
15 Weitzel G (1971) Therapie mit schnellen Elektronen. In: Vieten H, Wachsmann F (ed) Handbuch der medizinischen Radiologie, Bd XVI/2, Springer-Verlag, Berlin Heidelberg New York, pp 1-67
16 Wideröe R (1977) Elektronentherapie. Strahlentherapie 153:133

17 Zuppinger A, Poretti G, Zimmerli B (1964) Ergebnisse der medizinischen Strahlenforschung, Bd I. Thieme, Stuttgart, pp 347-405
18 Zuppinger A, Poretti G (1965) Elektronensymposium. Montreux. Springer Berlin Heidelberg New York
19 Zuppinger A (1967) Treatment by supervoltage machines — electron beam therapy. In: Deeley TS (ed) Modern trends in radiotherapy 1. Butterworth, London, pp 250-259

Electron Therapy for Cutaneous Epitheliomas

H. Pourquier

A Sagittaire linear accelerator has been used for treating cutaneous epitheliomas at the Center of Struggle Against Cancer, Montpellier, since 1969. A total of 130 cases with primary or recurrent tumors were treated with electron beam therapy between 1969 and 1973.

Electron beam therapy was applied to primary or recurrent epitheliomas which were at least 2 cm in diameter. Most lesions were located in the facial areas, i.e., ala of the nose, orbital edges, and lips. Some lesions extended to involve the mucosa. Previously these tumors were treated by low voltage X-rays at 120 kVp. Over the years low voltage X-rays have been progressively replaced by electron beam therapy.

It is essential to have a thorough knowledge of the physical, radiobiologic, and dosimetry aspects of electron irradiation prior to its application for therapy. Accelerated electrons possess certain physical characteristics which are particularly suitable for treating superficial lesions such as cutaneous epitheliomas. The radiation of an electron beam falls off rapidly with depth and the depth of penetration can be regulated by varying the energy of the electrons. One may choose an energy between 3 and 16 MeV, depending on the size and the depth of the lesion. Isodose distribution must be obtained prior to the initiation of the therapy, in order to assure that the tumor receives the maximum dose with sparing of the underlying normal tissues. The surface dose is about 90% of the maximum dose, and therefore, there is some degree of protection of the normal surface tissue. This improved dose distribution is responsible for fewer late complications, as compared with low voltage X-rays delivering similar doses.

Treatment factors are chosen according to the clinical parameters such as the size of the lesion in three dimensions, the amount of infiltration and the duration of the lesions. The margins used to cover the lesion depend on the tumor size, e.g., a few millimeters for the smallest and 1 cm or more for larger lesions. The selection of appropriate energy of electrons is important. Experience has shown that central recurrences are extremely rare when the whole lesion was included in the 80% or 90% of the isodose line. The radiation dose must be adequate. Cutaneous epitheliomas, regardless of the histological type, are radioresponsive. They can be sterilized by doses of 6000-6500 rads, delivered over a period of 5-6 weeks. Fractionation of dose is known to influence the biologic effects. Most patients in this series were treated with a total tumor dose of 6000-6300 rads by giving 400-430 rads per treatment, three times a week (NSD 2100-2300 rets). Close to the end of treatment the dose per session was reduced to 350-380 rads. In areas where the tissues do not tolerate treatment well (ears and temples) a smaller daily fraction of 200-225 rads was used. The advantages of individualized, careful selection of treatment factors are evident, particularly in the protection of the normal tissues in the facial areas. Experience has shown that favorable results may be obtained in structures involving cartilage such as the nose and ears. Electron beam therapy is superior to curie therapy because electrons can treat lesions that are more extensive. Electron beam therapy is also preferred to surgery because surgery will leave scars and will also delay the radiation therapy.

Acute radiation reactions after treatment usually subside spontaneously within a few weeks. Healing of radiation dermatitis may be accelerated by applying protective bandages. The long term late consequences observed include some cases of cutaneous dyschromia and fibrosis which usually occurred after extended irradiation. A comparison of late radiation sequalae, from irradiation by contact X-ray, superficial low voltage X-ray, and electron beam therapy has been made and is shown in Table 1. It can be seen that the incidence of severe dystrophy is less than 8%. However, it must be pointed out that 20%-30% of the cases treated by electron beam were recurrences after previous treatments. In view of this fact, an incidence of 7% late complications can be considered low.

Table 1. Cutaneous epitheliomas — dystrophic consequences after radiotherapy (1964-1973)

	Patients	Accidents, temporary necrosis	Severe definitive dystrophies
X-rays contact	708	31 (4.5%)	66 (0.8%)
X-rays superficial	222	36 (16.2%)	15 (6.7%)
Electrons (4 years)	129	19 (14.7%)	10 (7.7%)
Total	1059	86 (8.1%)	31 (2.9%)

Results (Figs. 1-3)

The results of treatment were analyzed. In the primary previously untreated group there were 37 patients with T_2N_0 lesions who had adequate follow up. Of these, 35 were free of disease at 5 years. The control rate was 95%. In patients with T_3N_0 lesions (11) local control was achieved in seven of eight patients (87%).

In recurrent cases, 43 patients were treated. Of these, 40 were cured at 5 years. However, there were nine late complications (four of which were severe) five of which were cured. A female patient whose disease reactivated was reirradiated, and cured.

Electron beam therapy offers excellent dose distribution, especially suitable for the treatment of cutaneous epitheliomas. Electron therapy is given basically to extensive tumors or recurrent tumors on the face, particularly in critically located areas which usually do not tolerate low voltage X-ray therapy. The 5 year cure rate of T_2 lesions was about 90%. The current rate was 80%-85% for T_3 lesions. Some lesions were dif-

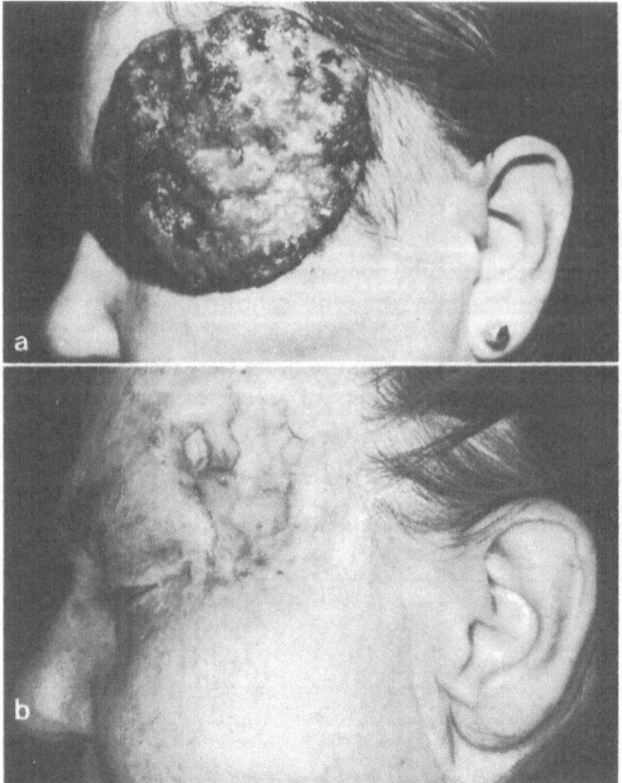

Fig. 1. a Spinocellular epithelioma together with a pre-auricular N_1 adenopathy treated by 13 MeV electron therapy through a bolus: 6200 rads in 5 1/2 weeks. Surgical procedure after a moderate telecobalt irradiation (3500 rads) on the said adenopathy. **b** The same patient 5 years later

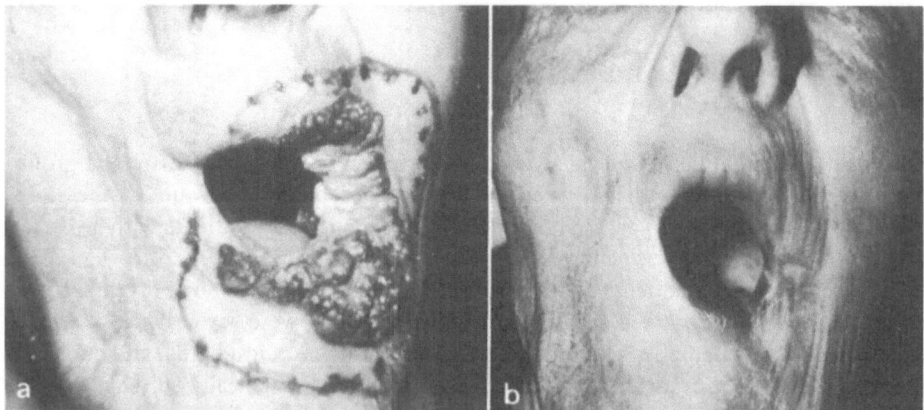

Fig. 2. a A mucous cutaneous spinocellular epithelioma covering the whole lip commissure and over 2 cm of the cheek internal wall. The upper and lower lip have been highly infiltrated, T_3N_0. Electrontherapy at 13 MeV: 6000 rads in 6 weeks by applying three treatments a week. First sessions at a rate of 430 rads, and 380 rads for the last ones. **b** The same patient 5 years later

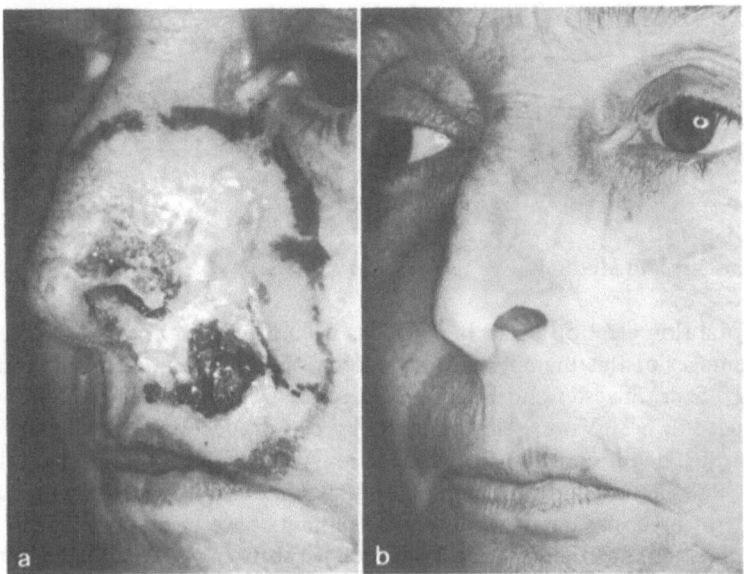

Fig. 3. a A patient showing an epithelioma at the nostril and internal nose treated by a dose of 8 MeV electron therapy at a rate of 1200 rads a week to a total dose of 6100 rads in 6 1/2 weeks. **b** The results 3 years later

ficult to sterilize, mainly because of the appearance of lymph node materials. Some lesions which showed treatment failure due to marginal recurrence were cured by surgical excisions.

Summary

The results of electron beam therapy were superior to other methods of treatment. The functional and cosmetic results, especially in the facial areas, were far better than conventional treatments. This is particularly true when appropriate electron therapy technique is applied to a previously untreated tumor.

Electron Beam Therapy in England

M.F. Spittle*

Patients and Methods

Total skin electron beam therapy has been available in London since 1962. The advantages of this therapy have been described by Trump et al. [1] and Bagshaw et al. [2]. Szur, Silvester and Bewley [3] described the method by which the electrons produced at 8 MeV were slowed down by carbon decelerators to produce a beam with an 80% depth dose at 4.5 mm or 6.5 mm. The AE1 6 MeV linear accelerator installed at the Hammersmith Hospital in 1969 was modified to treat skin lymphomas, particularly mycosis fungoides. Over 100 patients have been treated with this newer machine, most of them having been referred from my collegue Dr. Samman at the joint clinic at St. John's Hospital for Diseases of the Skin [4]. The pencil beam of radiation produced is diffused by passing it through a brass scatterer. This method was found to produce less X-ray contamination than the use of the gold scatterer [5]. Although X-ray contamination of the beam is 6% within the treatment cone, the patient is centred outside this X-ray field and the contamination is, therefore, reduced to less than 1%.

Although electrons are produced at 7 MeV the equivalent of approximately 0.5 MeV energy is absorbed in the interposed air between the patient and the beam. Three decelerators, two carbon and one copper, have been devised to reduce the energy of the beam further to effective voltages of 2.5, 3 and 3.5 MeV. The 7 mm carbon decelerator produces a beam with the 80% depth dose at 5.5 mm, the 4 mm carbon decelerated beam reaches the 80% at 8 mm and a 0.3 mm copper decelerator produces electrons with the 80% at 11.5 mm (A. Alderton 1976, personal communication). In recent years the more penetrating beams have been most frequently employed.

As the patient stands to one side of the main axis of the beam there is a dose gradient across the field and dosimeters on each side of the treatment area record a 7.5% fall-off across this region. The patient is rotated towards the source of electrons rather than facing down the main axis of the room. The isodose lines are vertical to a level of 2.5 m, which is adequate for treatment. The machine is set to irradiate horizontally and the patient stands 6 m from the machine displaced laterally by 1.3 m from the main axis of the X-ray beam. Isodose perturbation at ground level is avoided by placing the patient on an 8 cm platform. A disposable gown is worn and lead glass goggles are added if the head is also to be treated. To protect the testes an 11 cm wax shield

* My thanks are due to Dr. Peter D. Samman at St. John's Hospital for Diseases of the Skin and my predecessor the late Dr. Leon Szur who have allowed me to become interested in this group of patients. I would also like to thank Mrs. A. Beck for her help in treatment, Mr. David Hare for his advice and Mrs. Judith Leach for secretarial assistance

is used, this is lined with 3 mm of lead to absorb the bremsstrahlung from low atomic number material. If the head is not to be treated a 15 mm thick perspex shield is placed at the appropriate level. Any shielding of other parts, e.g. previously heavily treated small areas, may be undertaken using especially tailored block board.

The patient is treated four times weekly. On days 1 and 3 both an anterior and a posterior field are treated and on days 2 and 4 right and left lateral fields are treated. On each occasion 200 rads given dose is administered and this is recorded as 200 rads total when all four fields have been treated once. Initially the treatment doses achieved were in the order of 1000 rads and gradually this dose has been increased, patients receiving up to 2600 rads in ten treatments over 5-7 weeks. Lithium fluoride dosimetry has shown that such is the extent of overlap of these fields that when the anterior, posterior and right and left lateral fields have each been treated once with a given dose of 200 rads approximately 400 rads is received by most of the skin. It is important to note that at the Hammersmith Hospital the patient is recorded as receiving 200 rads when each of the four fields have been treated once to 200 rads. Therefore, the total dose expressed for a full course of X-ray treatment ranges between 1600 rads and 2600 rads when each area of skin has in fact received approximately double this dose.

It is inevitable that when treating the whole of the skin some area will be shielded from the beam is some treatment positions. Topping up doses to the axillae, the under surfaces of the feet and the perineum are frequently needed to ensure uniformity. This can be given with X-rays in the 80-100 kV range. The epiphora which is commonly due to infiltration of the eyelids with mycosis fungoides can be treated by using an internal lead eye shield with superficial X-rays given to the areas covered by the lead spectacles used for the electron beam therapy.

On completion of treatment generalized erythema and desquamation is seen. This varies in severity between patients and is most severe in those exhibiting extensive disease initially. In these patients it is often impossible to achieve the full dose and protracted intervals in the treatment time may be necessary. Pigmentation, ankle oedema and gynaecomastia are frequently seen. Fingernails and toenails are always shed 1-3 months following therapy and good regrowth of the nails is seen. Temporary alopecia occurs when the head is treated. Where follicular mucinosis has destroyed the hair follicles, patches of permanent alopecia remain. Bilateral brawny ankle oedema may be troublesome in older patients particularly if a lymphangiogram has been recently performed.

In other centres total skin electron beam therapy has been attempted using different modifications of the linear accelerator. At the Middlesex Hospital Planskoy (B. Planskoy 1978, personal communication) has devised a two field technique treating fields 42 x 42 cm with scattered 5 MeV electrons. The patient is placed at an angle of 10% to the horizontal, thus receiving an increased dose in the lateral regions. An electron beam is produced with an 80% depth dose at 13 mm and if a gap of 8 cm is left between the two large fields a homogeneous dose distribution is achieved across the anterior surface of the patient. To reduce the depth dose a cotton sheet may be placed over the patient.

In Manchester the patient is placed on a treatment couch which moves under the electron beam on tracks and again good uniformity of dose is achieved. The treatment may be fractionated as convenient locally and 600 rads once weekly may be repeated on two occasions and achieves useful disease response in patients with advanced disease.

Since 1964 an apparatus utilizing the mono-energetic 2.2 MeV electron beta-ray emitted from strontium 90 has been utilized to treat patients with very superficial mycosis fungoides [6]. Unfortunately the 90% depth dose is as superficial as 1 mm and, therefore, only early disease may be treated and the duration of remissions is not long. The strip of strontium 90 traverses the length of the carriage giving 50 rads to the patient's skin on each traverse. Two hundred rads are given on each occasion to an anterior, posterior and right and left lateral surface and the treatment is repeated until 2000 rads is given in ten treatments over 12 days [7]. The extreme simplicity of this unit has much to recommend it, but as with all the other methods of irradiating the whole skin there is considerable lack of uniformity.

Results

At St. John's Hospital for Diseases of the Skin we have found the long-term effects of electron beam therapy for mycosis fungoides less impressive than the results reported from Stanford [8], although we have some patients in remission for more than 7 years. We have tended to treat the more advanced cases where the average duration of remission following electron beam therapy has been approximately 18 months. The patients that have been treated with electron beam therapy for minimal disease have remained clearer for longer periods and there is certainly a dose related response. A patient in whom an overdose of electrons was received because of a mechanical fault achieved complete clearance of disease in the areas which received the overdose, whereas persistent disease remained in those areas that had had only minimal electron beam therapy.

While we have patients with tumor stage disease who have remained completely clear for over 2 years after electron beam therapy the disease usually recurs within 2-6 months after therapy in these advanced cases. In our experience it is unusual for a patient to be cured permanently by electron beam therapy although we do have two patients who have remained disease free for 10 years and who initially had widespread skin involvement. Although initial relapse following the treatment tends to occur in areas of low dose, e.g. the groin and axilla, the general recurrence which follows does not necessarily seem to spread from those areas but appears to be recrudescene of the disease in all sites.

At approximately 1 year following the completion of radical electron beam therapy many patients exhibit telangectasia, especially on the anterior surface of the trunk, the dorsum of the penis and the elbows. Although the cosmetic deformity is minimal we are anxious that these patients may have impaired long-term functional capability of the skin. One patient with extremely extensive disease developed small painful necrotic ulcers over the whole of the anterior trunk. This occurred at the same time as recurrence of the disease to which the patient rapidly succumbed. The dose achieved in these patients seems to be the maximum tolerated by the whole skin when this fractionation schedule is used.

We are not impressed by the high incidence of nodal involvement in mycosis fungoides although our policy has been to perform routine lymph node biopsies [9]. Approximately half of our patients eventually die from the accompaniments of skin ulceration and most of the rest from unrelated causes.

Recently the advent of long-wavelength irradiation in conjunction with 8-methoxy-psoralens therapy (PUVA) has revolutionized the treatment of most cases of mycosis fungoides [10]. Patients with superficial disease, no matter how extensive, are no longer given electron beam treatment until they have failed PUVA therapy. Again some areas do not receive adequate treatment from the ultraviolet and occasional non-ulcerated infiltrated dermal nodules appear. Local supplementary treatment with low dose superficial X-ray therapy is given. The periorbital regions shielded from PUVA may also be treated by low voltage X-rays using external eye shields.

It is our policy now to give PUVA therapy initially to all cases of superficial mycosis fungoides in view of the general well being of the patient attendant upon this form of treatment. Those areas which appear too deeply infiltrating to be reached by PUVA treatment are then topped up by local X-rays. Electron beam therapy is now reserved only for those patients presenting with tumor stage disease where PUVA is ineffective. In some patients initially cleared of advanced disease by electron beam therapy the remission has been prolonged using maintanence PUVA. This is now our policy with extensive disease. The disadvantages of chronic ultraviolet light therapy will probably include the induction of skin malignancy. When all the long-term effects have been adequately documented the place of photo-chemotherapy and electron beam therapy in the treatment of mycosis fungoides can be adequately evaluated.

Summary

Since 1962 total skin electron beam therapy has been available in London for the treatment of patients with mycosis fungoides and skin lymphomas. The 6 MeV linear accelerator produces a pencil beam of electrons at 7 MeV which are scattered through a brass scatterer and decelerated by either of two carbon decelerators or a brass decelerator to produce a beam of effective energy of 2.5, 3 and 3.5 MeV. These beams have an 80% isodose distribution at 5.5, 8.0 and 11.5 cm respectively. The patient receives 200 rads to an anterior, posterior, and right and left lateral field and achieves a total dose of between 1600 and 2600 rads in ten treatments in 5-7 weeks. Complete clearing of the disease can be predicted in all but patients having the most advanced tumours. However, the duration of remission after electron beam therapy is approximately 18 months and we are at present investigating the combination of psoralens and ultraviolet light therapy as maintenance treatment following lower dose electron beam therapy.

References

1 Trump JG, Wright KA, Evans WW, Anson JG, Hare HF, Frommer JL, Jacque G, Horne KW (1953) High energy electrons for the treatment of extensive superficial malignant lesions. Am J Roentgenol 69:623-629
2 Bagshaw MA, Scheidman HM, Farbar EM, Kaplan HS (1961) Electron beam therapy of mycosis fungoides. Calif Med 95:292-297

3 Szur L, Silvester JA, Bewley DK (1962) Treatment of the whole body surface with electrons. Lancet 72:1373-1377
4 Samman PD (1976) Mycosis fungoides and other cutaneous reticuloses. Clin Exp Derm 1:197
5 Spittle MF (1977) Mycosis fungoides. Electron beam therapy. Bull Cancer 64/2: 305-312
6 Haybittle JL (1964) A 24 curie strontium 90 unit for whole body superficial irradiation with beta rays. Br J Radiol 37:267
7 Bratherton DG (1972) Strontium beam therapy. In: Deeley (ed) Modern trends in radiotherapy. Butterworths, London, pp 176-187
8 Fuks ZY, Bagshaw MA, Farber EM (1973) Prognostic signs and the management of mycosis fungoides. Cancer 32:1385-1395
9 Spittle MF (1975) The treatment of mycosis fungoides. Trans St John's Hosp Derm Soc 61:31-34
10 Gilchrist BA, Parrish JA, Tanenbaum L, Haynes HA, Fitzpatrick TB (1976) Oral methoxsalen photochemotherapy of mycoses fungoides. Cancer 38:683-689

Fast Electrons in Treatment of Cancer of the Paranasal Sinuses

C. Jaulerry, J.P. Bataini, and F. Brunin

Carcinomas of the paranasal sinuses are uncommon tumors. Early cases with minimal or no bone destruction are usually treated by primary resection with or without post-operative irradiation. In advanced cases radical surgery is advocated, sometimes preceded by intra-arterial chemotherapy and/or radiotherapy, or it can be followed by radiotherapy.

At the Curie Institute, which is mainly a radiotherapy center, carcinoma of the paranasal sinuses is managed mainly by primary radiotherapy. Localized carcinoma of the anterior ethmoid usually adenoid-cystic or adenocarcinoma is an ideal indicating for electron beam therapy. Unfortunately early cases are often treated by surgery. Two cases treated by electron beam have been controlled for more than 10 years. Two very advanced cases failed locally. Extensive tumors of ethmoid, which are usually undifferentiated, are, however, best managed by wide field photon techniques as are also the extensive squamous cell tumors of the maxillary antrum, unless there is an important anterior, orbital, or malar extension when employment of electrons appeared preferable. Cancer of the ethmoid requires 20-25 MeV whereas for cancer of the antrum 30-35 MeV are employed thus ensuring theoretically 90% of the applied dose at a depth of 6-6.5 cm from the anterior plane and 80% at a depth of 8-8.5 cm.

Because of the diffusion of electrons in the tissues resulting from the interposition of the path of the beam of air cavities and bony septa, certain uncertainties exist as regard the absorbed dose in the tumor and surrounding tissues. Dose distribution was thus studied experimentally in a tissue-equivalent phantom moulded around a skeleton. The tissue dose in the antrum, the nasal fossa, and the pterygo-palatine region varies between 80% and 90% of the applied dose using a beam of 30 MeV.

However, in practice these discrepancies in dose distribution are certainly less important because of the filling of the air cavities and destruction of bony walls by disease. Nonetheless, because of the severity of the mucosal reaction mainly in the nasal fossa and the consequent synechiae, exclusive treatment with electron beam in management of carcinoma of the paranasal sinuses has been abandoned in favor of photon beams.

From 1962 though 1972, 26 patients with extensive carcinoma of the antrum were treated exclusively with fast electrons. In many cases bone destruction was important, involving all the antral walls and often the pterygo-palatine fossa. Spread of the disease to the nasal fossa and ethmoid was usual, and invasion of the floor of orbit common. The histology in almost all cases was squamous cell in type. Adenocarcinoma or adenoid cystic carcinoma was reported in two instances.

Irradiation Technique of Carcinoma of Maxillary Sinuses

A straight-on anterior open portal was employed covering the projection of the antrum, ethmoid, and nasal fossa. Commonest field sizes were 8 x 8 and 10 x 8. Part of the dose was sometimes given through a lateral portal. The energy level used depended upon the posterior extension of the disease. Usually nominal 30-35 MeV electrons were employed. Portal and energy levels were often reduced when the disease was mainly anterior (down to 25 MeV) for the last third or quarter of the treatment. Build up or compensators were not employed. Eye shielding was direct, using a lead plate 1.5 cm thick sandwiched between two layers of tissue equivalent material (0.5 cm thick). The electron transmission through this shield was nil and the dose due to bremsstrahlung was negligible. Tumor doses on the 90% isodose curve ranged from 5500 rads in 4 weeks to 7500 rads in 6 weeks delivered in five sessions a week.

As stated, mucosal reactions involving the mucosa of lip and cheek and especially the nasal mucosa were severe. Most severe skin reactions occurred around the labial commissure, the region of the inner canthus and free border of the lower lid.

Results

The absolute 3-year survival in this series was 9 out of 26 patients, similar to that obtained with photon techniques. The local controlrate at 2 years was 40%. There was no failure due to the disease between the third and the fifth years. These results compare favorably with the results of radical surgery for extensive disease. Failure is due to lack of local control of the lesion in almost all cases.

At the Curie Institute the role of electron beam therapy in the radiation management of cancer of the antrum and cancer of ethmoid unless very localized has now been reduced. Complementary or booster doses with electrons, however, sometimes prove useful.

Electron Therapy in Patients with Extensive Lip Cancer

R. Greiner and A. Zuppinger

Localized small lip cancers (T_1 and T_2) can easily be controlled either by surgery or irradiation. We prefer radiation because the cure rate is high and the functional results are very good. Surgery has only been applied in very small lesions which can readily be controlled by "enlarged" biopsy or in a few patients with a small local recurrence. The main problems are the T_3 and T_4 patients and the extensive recurrences where surgery is generally recommended. Jörgensen and co-workers, who have excellent results in T_1 and T_2 cases, recommend surgery in T_3 cases because they had a recurrence rate of 25% using irradiation alone. On the other hand, surgery is not possible in many cases because the tumor is often inoperable or the patient is too old, and frequently the patients refuse mutilating interventions or the operation risk is considered to be too high.

In 1961 we began treatment of wide spread lip cancers with electrons, initially in extensive recurrences after operation and irradiation because we were confronted with the question of survival. Since the first results were surprisingly good, we applied electrons in all T_3 and T_4 cases and in all recurrences with the same extent as the primary. In four patients we applied preoperative electron treatment, but we soon abandoned this procedure because of recurrences in two patients. We applied surgery only for small residual tumors or recurrences. This had to be performed in only three patients. All these remained cured. Fifty-one patients with T_3 and T_4 were treated with electrons between 1961 and 1972. All cases were epidermoid carcinomas, with the exception of one hemangioendothelioma and one basal cell carcinoma.

First, we applied direct fields, but we changed early to a tangential treatment because the isodoses were much better and it is easier to irradiate the angle of the lips which was clinically involved in 14 patients out of 51.

We introduced a wax covered lead mold in the vestibulum oris to protect the teeth, as far as they were present, or the bone, as far as it had not been involved (four patients). The fields were chosen only so large, so that in an eventual later necessary surgical procedure, the tissue would have to be removed in any case. The regional lymph nodes were clinically involved in 37% of patients. The submandibular region was generally included in the irradiation field of the primary tumor, in other cases we applied a prophylactic treatment. In those patients with clinical involvement of the submandibular lymph nodes, the parotid triangle was included in the regional field, followed by a submandibular lymph node resection.

At the beginning we had difficulties in the dosimetry, not only in lip cancer, but we soon recognized that the former recommendation of adding 10% was too small, especially in those patients in whom we applied higher energies. We had to make a virtue of necessity. We applied a biophysical method already inaugurated by Coutard. When the tumor regressed early and the normal tissue showed early reaction, we stopped the treatment before 6000 rads had been applied. In the contrary situation, we raised the dose until the normal tissues entered into a good reaction. This procedure has the advantage that the different sensitivities of normal tissues and that of the tumor are included in the dosimetry. The doses which lead to tumor sterilization varied from 5770 to 7960 rads respectively 1767 to 2330 rads. In case of early reaction of the normal tissues and insufficient tumor shrinkage, the treatment was stopped and re-treatment began after the recovery of the normal tissues. This split therapy was applied in 23% of the patients.

The results can be seen in Table 1. Only 15 patients were referred without previous treatment; all these tumors could be controlled (Fig. 1). In the 12 patients who were referred with a recurrence after operation or with incomplete surgery, all were cured. Only in one case was an excision of a residual tumor necessary.

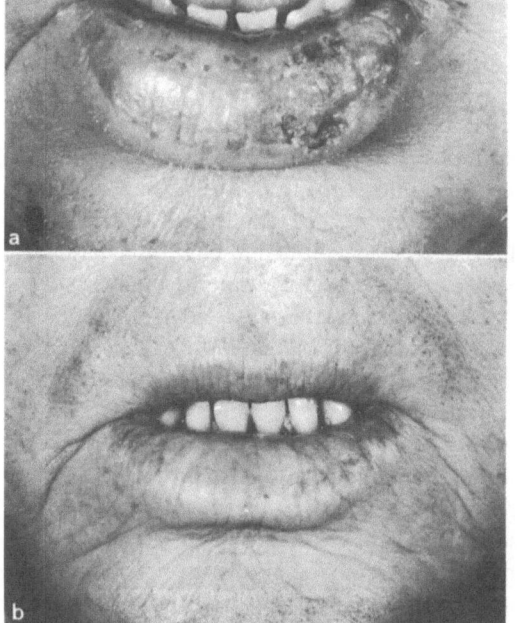

Fig. 1 a, b. 66-year-old patient with epidermoid cancer infiltrating the whole lower lip with superficial ulceration on the left side. T_3N_0. 30 MeV electrons 5770 rads in 56 days and 23 irradiations, 1750 rets. Split therapy because of strong early reaction, interval 4 weeks. a Patient before irradiation. b Patient 2 years later. No functional disturbances resulted, the patient was later active as priest, and was free of the disease after 13 years

Much more astonishing are the results in the group of recurrences after operation and irradiation or after irradiation alone. Out of 23 patients with curative treatment, we have had only three failures. One local failure occurred in the small series of the preoperative trial, and the only regional failure was due to the interruption of the treatment by the patient with a N_2 lymph node recurrence which reacted very well to the treatment. It is noteworthy that this regional recurrence occurred in a patient in whom the prophylactic treatment of the lymph nodes had been forgotten.

Table 1. Carcinoma of the lips, stage T_3, T_4 (1961-1970) 5 year survival rate [a]

	n	Pall. RT	5-year survival NED	Death due to other	Recurrence local plus LN	distant metastases
Recurrence after previous irradiation or operation + irradiation T_3 (15), T_4 (9)	24	1	18	2	2	1
Incomplete resection or recurrence after previous operation T_3	12	1	9	2	—	—
Irradiation only no previous treatment T_3 (10), T_4 (5)	15		11	4	—	—
	51	2	38	8	2	1

	All cases	Without patients of 80 and more years
Absolute 5-year cure rate	75%	80%
Relative cure rate	78%	82%
Recurrence rate	6%	4%

1 case is a hemangiopericytoma
1 case is a basal cell carcinoma

[a] Patients with fixation to the bone or with metastases fixed to the bone where classified as T_4.

PALL. RT, palliative radiotherapy only;
NED, no evidence of disease;
LN, lymph nodes

No severe damage occurred. Twice we saw a small sequester which could be cured by local excision. Few patients showed indurations, only one patient had a recurring superficial radium ulcer after radium needling, but he refused surgery.

Discussion

This series with tumors which are generally considered to be incurable or to have a very high recurrence rate after irradiation and surgery showed surprisingly good results with electrons. There was only one real failure of irradiation and this was in a very extensive inoperable tumor recurrence. Localized lip tumors (T_1 and T_2) can easily be treated by contact therapy with very good cure rates and functional results. In the case of recurrence, local excision can be done, and in the more extensive recurrences electron beam therapy has a very good chance of cure. In all T_3 and T_4 patients electron therapy should be applied. We do not agree with the radiotherapist Fletcher who recommends only surgery in these patients. We have no experience in patients with radiologically visible bone involvement. Prophylactic irradiation of the lymph nodes is harmless and necessary because the prognosis of patients with positive lymph nodes is poor, with only about 30% survivals in 10 years. The very good prognosis of lip cancer with irradiation alone is not known enough by surgeons and practitioners.

Summary

Extensive lip cancers (T_3 and T_4) are a very good indication for high speed electron therapy because it offers a very high cure rate and good functional results with low risk of damage. Recurrences after operation and/or irradiation have a very good chance of control with this therapy, which is much better than extensive surgery. Surgery is necessary only in a few patients with local and regional recurrences. (More details are available in a thesis of the Medical Faculty of Berne by Vroni Schneeberger.) Here also the whole literature can be required.

Tumors of the Salivary Glands

A. Zuppinger

This tumor group — the majority arising from the parotid gland — is very interesting from the tumor biologic standpoint on account of the manifold histological variabilities with different clinical behavior which should be considered when planning treatment. Mixed tumors, mucoepidermoid and adenocystic carcinomas, which were formerly often taken together as semimalignant tumors, were considered to be radioresistant, so that only the other malignant tumors were sent for radiotherapy, and generally only after operation or recurrence after surgery. The results with conventional irradiation, contact therapy, and high voltage irradiation were not satisfactory. The therapy of these tumors with electrons, which was first recommended by Becker [2] mainly for physical reasons, has entirely changed the treatment and prognosis of these tumors, though for superficial cases the isodoses are similar to contact therapy.

We began this treatment in 1958 and had treated, up to the end of 1971, 114 tumors of the parotid (106) and submandibular (8) glands. Our good results are due not only to the electrons but also to our very good collaboration with the laryngologists Professors Escher and Neiger. We will discuss the various tumors separately because the procedure is different for each.

In benign mixed tumors, which formerly were considered to be entirely radioresistant, we recommend electron treatment only when radical surgery is not possible, as postoperative treatment, or when surgery is refused. Two females with recurrences after surgery refused another operation because they would run the risk of a facial lesion; they received only radiation and have lived with small stable residual tumors for 12 and 17 years, while the others are free of disease. In all cases no facial lesion occurred. We know that in many places this indication is refused; in Berne also the great majority of cases are only operated on. But the good results in nonradical operation and recurrences prove that the tumor responds to electron irradiation. Irradiation in the above-mentioned situations has the advantage that the facial nerve can be spared and that *recurrence can be prevented with a high probability*. Relatively high doses — between 5000 and 6000 rads — which leave no, or only very slight, changes in the skin are necessary and do not cause complications in the event of a later operation.

Concerning *mucoepidermoid tumors*, which are considered by many physicians to be a true carcinoma, very little knowledge exists about their radiosensitivity except that they are considered to be radioresistant. Out of eight cases given curative treatment seven are free of disease. Three had postoperative irradiation, but two had no radical operation. In three out of four cases with recurrence after operation a residual tumor had to be operated on. For sterilization the tumors need doses of 6000 rads.

We do not force the cure only by radiotherapy; we prefer to give lower doses, 4000-5000 rads depending upon the shrinking, and have operated on a possible residual tumor. This policy of combined treatment enables us to reduce the damage to normal structures. It is possible because the operative procedure with these doses is not rendered more difficult; on the contrary, it is probably facilitated, since the tumor is smaller, and the patient benefits because the recurrence rate is lower. No patient in this series has a recurrence.

The above-mentioned policy is also used whenever possible in other salivary gland tumors, with the exception of highly sensitive tumors.

The most surprising effects were achieved in *adenocystic carcinomas*. We were encouraged to use electron treatment first by an inoperable ectopic adenocystic carcinoma of the hard palate in a 76-year-old man, which disappeared after split therapy. Out of 23 patients who received curative treatment 20 were locally free of disease (78% are free of disease after 5 years), but two had recurrences after 6 1/2 and 7 years. The technique in these two cases has to be considered retrospectively as wrong because we employed too low a voltage. Only four cases with inclusion of the above-mentioned recurrences developed *distant metastases.* This result is far better than other results in the literature. The reason is not only the better local control, but probably the operative procedure. Our surgeons know that the tumor is radiosensitive to electrons. When the tumor is not easily operable they only perform an excision or biopsy and send the patient for irradiation. This procedure probably reduces the operative spread in broad radical surgery.

In the other *malignant epithelial tumors* the results are not so good, but are much better than in former times. Forty-nine cases were sent for treatment; nine received only palliative treatment, seven of them because they already had distant metastases: all of them had formerly undergone incomplete surgery. Fifty percent of the patients given curative treatment are free of disease; four patients were locally free of disease but developed distant metastases: all had formerly had incomplete surgery. In 11 patients (27.5%) tumors could not be locally controlled. Seven of them had undergone a former nonradical operation and four, all with T_4 tumors, are radiation failures, mainly because the voltage was too low. The cases formerly irradiated will be discussed later on. There is no obvious dependence of the cure rate on the histology. It may be that epidermoid carcinomas can be more easily locally cured than the adenocarcinomas.

One hemangiopericytoma, a recurrence after several non radical operations, could be controlled with 6300 rads and the patient has been free of disease for 15 years.

Sarcomas have a bad prognosis. Out of seven cases only one is living after radical surgery and postoperative irradiation. All the others died after operation or biopsy and irradiation with electrons, though locally free of disease, because of distant metastases.

A survey of the whole tumor group shows that local control of salivary gland tumors is much easier with electrons than with conventional X-rays and high-energy photons. Often electron therapy in combination with surgery is the method of choice. The energy used must be high. We have lost at least four patients with a recurrence in the region of the processus pharyngeus of the parotid gland. The energy must be, in malignant tumors, at least 20-25 MeV.

Local control with 27% failures can easily be improved. The failures occurred in the cases with nonradical operation. We believe that preoperative irradiation at least in inoperable or borderline cases should raise the cure rate considerably. Surgeons' fears that the complication rate is high are no longer justified. This philosophy has in addition the probable advantage that the high rate of distant metastases can be diminished. The response to high-energy electrons in parotid tumors is so good that primary sacrifice of the facial nerve is no longer justified.

In malignant tumors the regional lymph nodes should be irradiated. Prophylactic neck dissection is not necessary.

We have also had some *radiation damage*. One case had a certain, and another a probable, damage of the cervical medulla. This can easily be prevented by directing the beam at an angle of 15° to the sagittal from behind to the anterior part of the skull, or by applying a pendular therapy which gives very good isodoses in the tumor region and a low dose at the medulla, with good protection of the skin. If the dose at the medulla exceeds 3000 rads we recommend applying the additional dose with a stationary field 15° from behind.

One patient developed a *bone necrosis* after a tooth extraction without protection; this was treated surgically by implanting bone from a rib, and healed without complication, thus proving that the tissue damage was not serious.

More details will be given in a paper which will soon be published [1].

In summary we can say that electron therapy has entirely changed the indication and results of salivary gland tumors. It is the treatment of choice. High energies of at least 20-25 MeV are necessary. Better results may be achieved, particularly with wide application of preoperative irradiation, which will most probably improve the local cure rate and diminish the high rate of distant metastases.

Summary

Our results with photon therapy have been disappointing. We began our treatment with high speed electrons in 1958. Up to 1971, 114 tumors of the parotid and submandibular salivary gland tumors have been treated, often in combination with operation; 93 were malignant, of them 79 received curative treatment. The mucoepidermoid carcinoma needs high doses. Out of eight cases treated curative seven are free of disease. In the adenocystic carcinoma there are only 5% local failures in the 5-year period. In the other carcinomas the 5-year cure rate is 50% and the local failure rate is 27.5%. In 12 cases with recurrences after operation and irradiation the cure rate was also 50%. Energies of 20-30 MeV are necessary because of the recurrences in the pharyngeal processus in applying lower energies. No facial lesion has been observed. In the benign mixed tumors electron therapy is only applied in cases with nonradical surgery or when surgery is refused. Out of 20 cases with electron therapy 18 are free of disease and two have had a stable residual tumor for 12 and 17 years respectively. Electron therapy, mostly in combination with surgery, is the radiologic treatment of choice.

References

1 Zuppinger A, Escher F (1979) Schnelle Elektronen bei der Therapie von Speicheldrüsentumoren. Strahlentherapie 155:75-81
2 Becker J, Schubert G (1961) Die Supervolttherapie. Thieme, Stuttgart, S 360

Importance of High Speed Electrons in the Treatment of Lymph Node Metastases

J. Bernier and J.P. Bataini

In the Curie Institute, the treatment of metastases of head and neck epidermoid carcinomas is essentially by radiotherapy. The clinical application of the electron beam of the BBC betatron since 1962 has strengthened this attitude. The importance of high speed electrons in the therapy of metastatic lymph nodes has been studied in 219 patients treated between 1970 and 1977 who had previous surgery or irradiation. In the same period 65 patients who had had previous surgical or radiologic treatment were treated by high speed electrons because of tumor recurrences.

Patients and Methods

The origins of the primary tumors which caused the metastases are seen in Table 1. The cancers of the oropharynx and hypopharynx prevail (43% and 26% respectively). The point of origin could not be found in 11.5% of the patients (25/219) and almost 8% had multiple primaries. Only one patient had no palpable lymph nodes and received only an elective regional treatment. Table 2 shows the distribution of the patients following the TN and the classification proposed by the American Joint Committee. The size of the lymph nodes can be seen in Table 3.

Table 1. Origin of the primary tumors of the 219 patients [a]

Oropharynx	95
Hypopharynx	58
Unknown	25
Buccal cavity	16
Rhinopharynx	11
Supraglottic	5
Lips	5
Paranasal sinuses	4

[a] 17 patients presented with multiple lesions in the head and neck region

Table 2. Clinical classification of 219 patients

UICC		AJC	
N_0	1	N_0	1
N_{1b}	60	N_1	29
N_{2b}	8	N_{2a}	56
N_3	150	N_{2b}	38
		N_{3a}	52
		N_{3b}	43
Total	219	Total	219

Table 3. Size of lymph nodes (218 patients N+)

Size of the adenopathy	
N ≤ 2 cm	63
3 cm ≤ N ≤ 5 cm	116
5 cm < N ≤ 7 cm	33
7 cm < N ≤ 10 cm	33
N > 10 cm	2
Total	247

Electron therapy is applied in three different modalities:
a) In additional irradiation of lymph node residuals after cobalt therapy, mainly in the posterior lymph nodes.
b) As the principial radiation modality when the target volume of the primary and the lymph nodes are separated as in nasopharyngeal cancer and in cancers of the paranasal sinuses and parotid.
c) In prophylactic treatment of the lymph nodes in the vicinity of the spine.

The adequately chosen energy spares the critical organs in the neighborhood, especially the spinal cord. As the minimal energy is about 10.5 MeV, the use of a bolus is often necessary. The dose is calculated on the 90% isodose curve. In the case of boost doses (additional treatment) the field size generally is of 15-35 cm^2. In the case of irradiation of the whole cervical region, the fields are enlarged to 80-100 cm^2.

The applied dose of cobalt therapy in the first treatment varied in the majority of cases between 6000 and 8000 rads. These doses (Fig. 1) were applied to the cervical region by continuously reducing the field size before the boost dose with electrons was given to the rest of the nodes. But a certain group received only 4000-5000 rads by 60 Co. These are patients in whom the whole cervical region was treated, partly by 60 Co and partly by electrons.

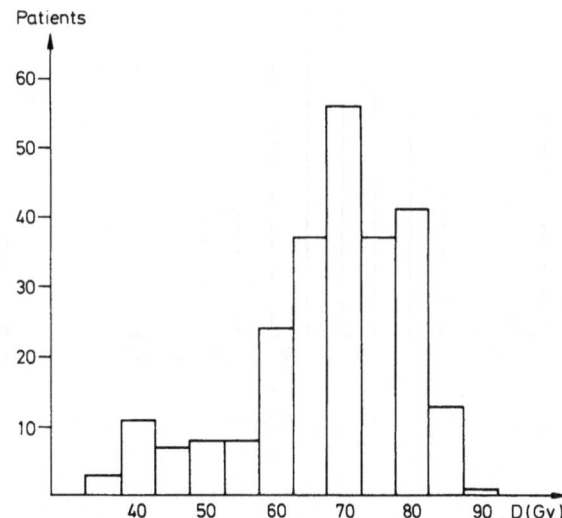

Fig. 1. Distribution of doses delivered to the neck with ^{60}Co

The supplementary electron beam doses varied from 500 to 3000 rads at the 90% reference isodose curve (Fig. 2). The total dose of ^{60}Co plus electrons was, in the majority, between 7000 and 9000 rads, depending upon the topography, the form, mobility, and the clinical behavior of the lymph nodes (Fig. 3). These high doses were only applied to the smaller fields, e.g., 4-5 cm at the end of the treatment of the remaining tumor residue.

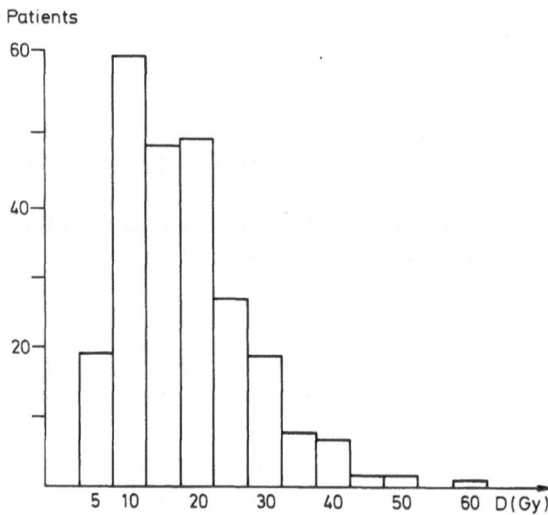

Fig. 2. Distribution of doses delivered with electrons

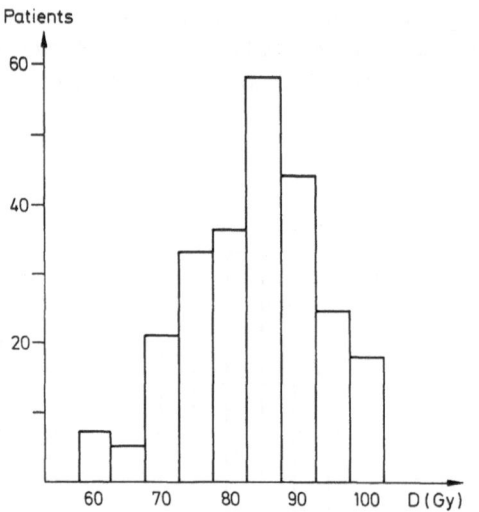

Fig. 3. Distribution of total doses delivered at the level of the lymph nodes using the combination of photons and electrons

Results

The global results are shown in Table 4 with an observation time between 1 and 7 years. Control of the lymph nodes was achieved in 65% (103/160). An isolated failure in the lymph nodes occurred in 22.5% (36) and a failure in the primary as well in 13% (21 patients). The appearance of the lymph node recurrences is relatively early: 65% arise before the end of the 1st year, 87% before the end of the 2nd year. Out of the patients with lymph node recurrences, ten had surgery with four successful results.

The lymph node control, depending upon the volume of the nodes, is analyzed in Table 5: 78% success in nodes smaller than 5 cm and almost 50% in those larger than 5 cm.

Table 4. Treatment with electrons: results

	Total results	Results after 2 years
Lymph node control	156/219 (71%)	103/160 (64.5%)
Failure of lymph node control	40/219 (18%)	36/160 (22.5%)
Local and regional failure	23/219 (11%)	21/160 (13%)

Table 5. Results depending on size of the nodes (218 patients)

Size of the lymph node	Number of patients (n)	Lymph node control	Lymph node failure	Local-regional failure
N ⩽ 5 cm	161	126 (78%)	21 (13%)	14 (8%)
5 cm < N ⩽ 7 cm	28	15 (53%)	6 (22%)	7 (25%)
N > 7 cm	29	14 (48%)	13 (45%)	2 (7%)

Table 6 shows that where there has been only a single node smaller than 5 cm, the success is 85% and almost 60% in multiple homolateral lymph nodes, nodes larger than 5 cm, and even with bilateral nodes.

Finally, the prophylactic irradiation of clinically noninvolved nodes of one or both sides was applied in 19 patients. No patient irradiated with ^{60}Co and electrons had a recurrence, but one developed a contralateral submandibular lymph node metastasis.

Complications

The early reactions appear more frequently and intensively when the treatment of the cervical region is carried out only with electrons. There may result a strong exu-

Table 6: Success rate related to size, number, and type of lymph nodes (218 patients)

	Number of patients (n)	Lymph node control	Lymph node failure	Local-regional failure
Single lymph node less than 5 cm	86	73 (85%)	9 (10%)	4 (5%)
Multiple homolateral or single lymph nodes larger than 5 cm	89	51 (57%)	21 (24%)	17 (19%)
Bilateral metastases	43	27 (63%)	10 (23%)	6 (14%)
Total	218	151	40	27

dative reaction necessitating an interruption of several days. This is the reason – even in those patients in whom the treatment of the primary tumor and the nodes can be dissociated – why we prefer a combination of photons and electrons, which proved to be better tolerated at the surface.

Severe late complications arise less frequently. Out of 169 patients who did not show signs of recurrence in the lymph node region, only 5 (3%) showed a severe sclerosis in the cervical region with functional impairment. Atrophy and depigmentation is more frequently seen than a real sclerodermic reaction. In addition, no radiation myelitis occurred in spite of a frequent involvement of the posterior cervical nodes and the prophylactic treatment of the spinal chain of nodes.

Discussion

The results achieved here demonstrate that a primary exclusive irradiation may result in the sterilization of involved nodes of the neck, even in patients with extensive involvement. To get such results the doses delivered at the site of the tumor has to be high. It is thus necessary to reduce the integral dose given to the surrounding tissues with the goal of preventing severe late complications.

It is for this reason that the application of high speed electrons in association with photons allows the delivery of a homogeneous dose in the superficial regions where the nodes occur. In addition, the neighboring highly radiosensitive structures can be spared. On the other hand, it is indispensable to progressively reduce the field size in relation to the response of the tumor to the irradiation. The choice of the radiation energy is important and often has to be diminished to the last part of the treatment.

Treatment of 65 patients with Lymph Node Recurrences with High Speed Electrons

This secondary irradiation was given mainly after a first exclusive irradiation had been applied (55/65). Six other patients had had surgery and four patients had had neither an irradiation nor surgery because they had been classified as N_0 at the beginning.

The results of the secondary irradiation (Table 7) vary, obviously depending upon the topography of the recurrence, where it is situated in the irradiation field (8 failures

Table 7. Results in 65 patients with lymph node recurrences

Topography of the recurrence	Number of patients (n)	Lymph node control	Lymph node failure	Local failure	Local-regional failure
In not irra-diated re-gion	30	16	8	6	–
In the bor-derline	24	5	16	–	3
In irradia-ted region	11	3	8	–	–
Total	65	24	32	6	3

out of 11), at the border of the field (16 failures out of 24), or in a region not previously irradiated (8 failures out of 30). Consequently, one out of two patients benefited by this irradiation when the recurrence developed outside the irradiated field, but lymph node control could only be achieved in less than one out of four patients with recurrent tumor in the irradiated field or at the border. Here the minimal follow-up time after the secondary treatment was a year.

Conclusions

An combination of photons and electrons has been applied in 219 patients with cervical metastatic disease. This treatment modality allows the delivery of high radiation doses in the region of the lymph nodes and protects deeper situated regions because of the rapid decrease of the absorbed dose at the intended depth.

Lymph node control could be achieved in 70% of all patients, though there was a high frequency of fixed lymph nodes and those with large size. Booster doses of electrons after ^{60}CO for residual nodes (especially when posteriorly seated), wide irradiation in association with ^{60}CO and, finally, elective or curative irradiation of the spinal chain of nodes are indicated for the application of electrons of appropriate energy in the management of cervical metastatic nodes. The voltage and the dimensions of the fields have to be modified during the treatment so that it is possible to apply a high dose only in the region of the involved lymph nodes. The selection of these therapeutic parameters is dependent upon the clinical evolution and on the experience of the radiotherapist with neck cancers.

Summary

The authors give a review of their results achieved with the combination of high speed electrons with photons in the treatment of lymph nodes due to epitheliomas of the head and the neck.

Between 1970 and 1977, 219 patients were treated by primary exclusive radiotherapy. The doses delivered by the combination of high speed electrons and cobalt varied between 7000 and 9500 rads by progressively reduced field sizes.

Two years after treatment 64.5% (103/160) of the patients were free of disease in the lymph node region of the neck. The failure rate in the lymph nodes was 22.5% (36 patients).

In addition, 65 other patients with recurrences in the lymph node region received a secondary irradiation with electrons. The results vary to the topography of the recurrences in relation to the regions formerly irradiated.

Electron Therapy as a Postoperative Modality for Head and Neck Cancers

F. Eschwege, P. Wibault, J.M. Richard, and C. Haie

Electron therapy, as a postoperative medium for head and neck cancers, has been used for many years at the Gustave-Roussy Institute in Villejuif. This therapy became possible after an Almis-Chalmers betatron was installed in 1954 and was greatly improved when a Sagittaire linear accelerator was also installed 1972.

The use of electron therapy is not based on any particular biologic argument, but it is based on the advantages and conveniences which can be understood through knowing the depth dose characteristics of electrons if different energies.

The advantages to be expected through its use are: possibility of maximum doses being delivered to the surface layers; sharing of contralateral normal structures; and the simplicity of its use. These practical advantage are particularly interesting when irradiating postoperative cases.

The *inconveniences* are not negligible. The frequency of errors when calculating depths, as well as the lack of knowledge concerning the transmission of electrons through heterogeneous tissues, involve risks of over or underdosing the patient; the variation may reach from 10% to 15% of the basic dose.

There are two other difficulties to consider: *First,* the particular way electron isodoses "mushroom" explains the problems that can emerge when a cobalt field is combined with an adjacent electron field. *Second,* overdoses on the surface layers, particularly when the doses are improperly fractionated, may create painful subcutaneous fibrosis in some cases, involving the postoperative neck.

All these difficulties should not be forgotten in daily practice, however, so far they have not made us decrease the importance of electron therapy in the postoperative treatment of head and neck cancers.

Postoperative Irradiation Modalities

Electron therapy may be used by itself for: treatment of tumors of the salivary glands (parotid and submaxillary), treatment of nose and ethmoidal tumors, treatment of the outer and middle ear tumors, and after cervical lymph node dissection.

We also combine it with cobalt after transmaxillary buccopharyngectomy, mandibulectomy, node dissection, exeresis surgery on maxillary sinus tumors, and transmaxillary pharyngectomy and laryngectomy involving cervical lymph node dissection.

The postoperative doses we deliver — by combining electrons with cobalt or by electrons alone — are as follows: for primary tumors 50 Gy if the resection was adequate, and 65 Gy if the resection was incomplete; for lymph nodes 50 Gy if no node was invaded histologically (N^-), or if only one or two nodes were invaded without capsular rupture (N^+R^-). If nodes were invaded with capsular rupture, we give 65 Gy.

Illustrative Cases of Postoperative Irradiation Using Electrons

Where *malignant tumors of the parotid* are concerned (mucoepidermoidal tumors), particularly after exeresis surgery, postoperative treatments are performed exclusively by means of electrons. After checking the histological and operative reports, the patient is placed according to the treatment position. Contours are taken in order to determine accurately the maximum depth to which the irradiation is going to be given. Then the electron energy is chosen (between 10 and 13 MeV). In general the beam is directed perpendicular to the skin. It is understood that the scar is to be covered. Where cutaneous irradiation is required, a bolus a few centimeters thick, should be used.

For tumors of the outer ear, until 1972 we used to irradiate with ^{60}Co by means of two oblique fields with wedges. Although the dosimetric study may seem satisfactory, this technique is difficult since the position of the fields has to be identical during all the treatment. We are actually using an easier technique by irradiating most of times with 8 MeV electrons through a direct field. Here the dose is satisfactory and the technique easier to reproduce.

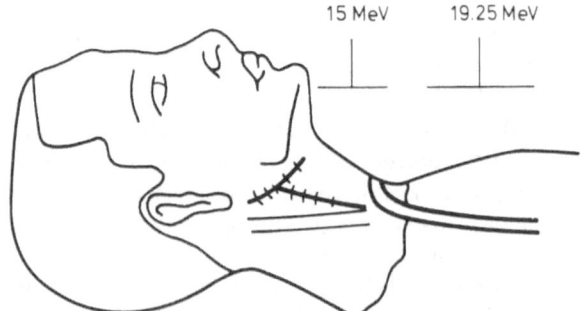

Fig. 1

For *tumors of the paranasal sinuses,* irradiation is performed systematically after exeresis surgery. This treatment may use only electron therapy, but more frequently electrons are used in combination with high energy photons.

For *tumors of the oropharynx after bucopharyngectomy,* electrons are used more frequently in combination with photons to underdose the contralateral side and, at the same time to give an adequate dose to the area of resection of nodes with capsular rupture.

The use of electrons for *pharyngolarynx cancers* has been systematic since 1972. Previously postoperative treatments were performed only with ^{60}Co. The patient was treated in the decubitus position by using posterior oblique fields, which were reduced systematically so as to avoid the spinal cord. This irradiation involved a considerable dose at the level of the tangential cutaneous entries and exits. There was the appearance of swollen areas under the chin showing subcutaneous fibrosis. Since 1972 we have been using an easier technique. As a first stage we irradiated with opposed parallel beams of ^{60}Co. The second stage involves reducing the beams so as to protect the spinal cord. The posterior area is thus treated with electrons whereas the anterior one is treated with cobalt.

Using such a technique means placing the patient under different apparatus following the same positioning and the treatment is to be carried out by the same team of professionals. These techniques have helped us to prevent the appearance of swellings under the chin, and in most cases led to the absence of fibrosis.

The use of electrons as postoperative therapy has been routine in cases of infiltrated nodes or in cases of pericanicular or tracheal recurrence (Fig. 1).

To end with, the use of electron irradiation for postoperative treatment of head and neck cancers has become a standard way of management at the Gustave-Roussy Institute, and electron therapy certainly has simplified our treatments, avoiding complicated photon beam techniques which are sometimes difficult to reproduce. This does not mean to ignore the theoretical problems mentioned previously and the necessary knowledge of the mode of tumor spread. It means that each case has to be analyzed thoroughly according to the clinical, surgical, and histological findings. All of these will give us the necessary information to determine whether it is advisable to use electrons alone or to use electrons in combination with photons.

The Place of Electron Therapy in Curative Treatment of Operable Breast Cancer

R. Amalric, F. Robert, C. Altschuler, and J.M. Spitalier

Electron beams have not been used alone (at the Cancer Institute, Marseilles) in irradiation of operable breast cancers, but we have been systematically using them for 10 years for additional doses to supplement target volumes following basic irradiation either with telecesium 137 or with telecobalt 60. We have thus treated 2626 cases of breast carcinoma since 1968.

Our technique is nearly the same as that described by Bataini (1970). Breast irradiation itself is performed in the lateral decubitus by means of two large tangential fields which cover the mammary gland; the dose of telecobalt is 50 Gy in 5 weeks. Immediately after this basic irradiation, an additional dose is given to the palpable remaining tumor or to the initial site of tumor (in cases of previous lumpectomy) with a 10-15 MeV electron beam (according to the thickness of target volume which will have to be included in isodose 90%) from a small direct field (Fig. 1). This complementary irradiation will give 30 Gy in 3 weeks, which brings the tumor dose to 80 Gy in 8 weeks.

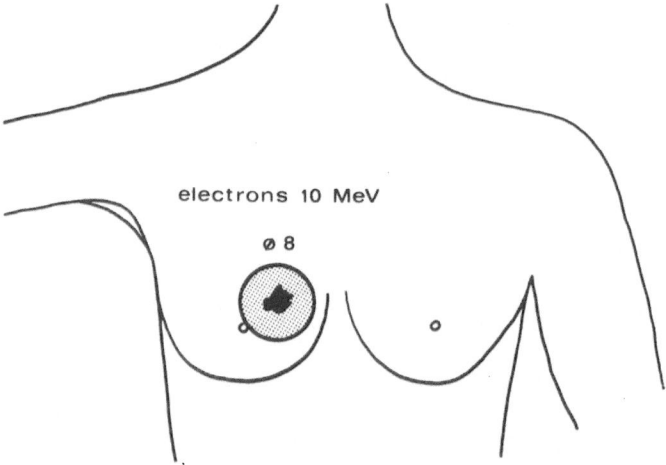

electrons 10 MeV

ø 8

Fig. 1. Additional dose on a second target volume using a direct field of 10-15 MeV electron beam

Irradiation of nodal areas is achieved as follows (Fig. 2):

1) For the *axilla*, basic irradiation is performed with telecobalt using two opposite fields (anterior axillary and posterior axillary) up to the dose of 50 Gy in 5 weeks. An additional dose is then given with 12-15 MeV electron beam from a small direct axillary field. The complementary dose is 20 Gy in 2 weeks in cases of palpable nodes; 10 Gy only are delivered in the absence of palpable nodes.

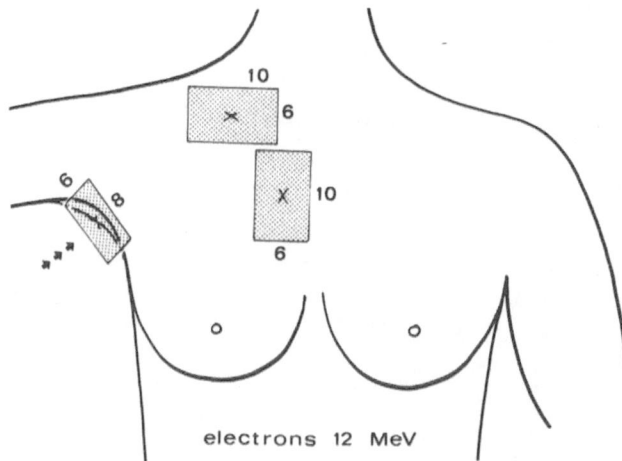

Fig. 2. Additional dose on node areas (axillary, supraclavicular, internal mammary) with direct fields of 12 MeV electron beam

2) For *supraclavicular* basic irradiation, 45 Gy are delivered with telecobalt in 4 1/2 weeks, from a single direct field. If palpable axillary adenopathy exists an additional dose will be given using a small direct field with 12 MeV electron beam and a 10 Gy dose in 1 week.
3) For the *internal mammary chain* on which the average dose is 35 Gy in 5 weeks with telecobalt from tangential mammary fields, 12 MeV electron beam is provided in addition if the cancer is located within inner quadrants or in the subareolar area. This additional dose is delivered from one direct field covering the three first intercostal spaces and centered 3 cm away from the midline; generally the size of this field is 6 x 10 cm and complementary average dose is 15 Gy in 1 1/2 weeks.

On the whole the *total dose* (^{60}Co + electron beam) will provide in axillary areas: 70 Gy over 7 weeks in the middle part of axilla (for a depth of 3 cm), in the case of N_1; 60 Gy over 6 weeks in the case of N_0; 55 Gy over 5 1/2 weeks in the superclavicular fossa; 50 Gy over 6 weeks in the internal mammary chain (Fig. 3). These doses and fractionations generally allow good tolerance of ^{60}Co and electron beam. Nevertheless skin reactions are observed in one-third of the cases; it may be moist desquamation in the places where additional doses were delivered and more so in the axillary fossa; recovery from these reactions is fast and of good quality, however.
We closely studied functional and aesthetic late results, establishing frequency and importance of radiation sequelae in more than 300 cases with a 5-year minimum follow-up. In 24% of the cases no radiation sequelae occurred and the irradiated breast was exactly the same as the opposite one. In 66% of the cases there were slight or moderate radiation sequelae, such as pigmentation, telangiectasia, or limited sclerosis. In other words 90% of the cases came out without any important deformation of breast, without breast sclerosis, or limitation of shoulder motion. In the remaining 10% of the cases, important changes were observed, i.e., (a) in 8% mammary or axillary fibrosis, well marked, and with limitation of arm motion; (b) only 2% had serious

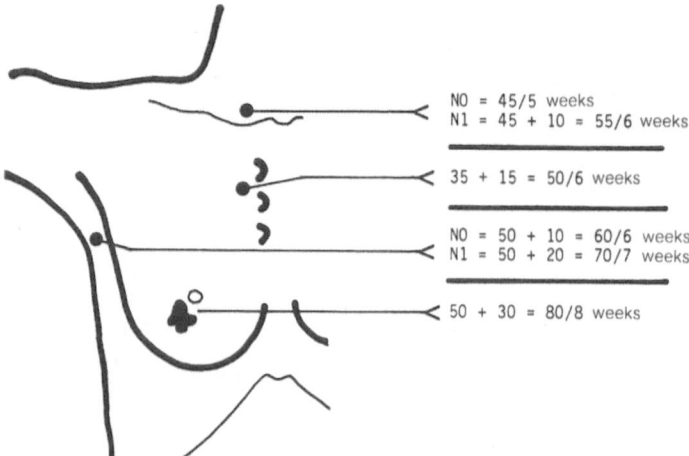

NO = 45/5 weeks
N1 = 45 + 10 = 55/6 weeks

35 + 15 = 50/6 weeks

NO = 50 + 10 = 60/6 weeks
N1 = 50 + 20 = 70/7 weeks

50 + 30 = 80/8 weeks

Fig. 3. Total doses of telecobalt + electron (Gy) for the tumor and node areas

complications involving invalidation (very big arm edema, injury of brachial plexus, etc.).

This low rate of complications is a good witness to the excellent tolerance of tissues to a combination of two-thirds gamma rays with one-third beta rays.

From the oncologic point of view the late results of this type of conservative treatment are at least as good as those of radical surgery:

1) In a 5-year follow up and for stages I and II without any sign of fast growth (T_1, T_2/N_0, N_1/no evidence of fast growth) we observe from 83% to 73% (whether lumpectomy has been done or not; over a total of 700 cases) free of disease.
   For ideal case $T_1 N_0$, the survival rate is 88% (with previous lumpectomy; 112 out of 127).

2) In a 10-year follow up and for stages both I and II gross survival percentages (uncorrected according to age; 52 cases) spread from 69% to 61% (whether lumpectomy has been done or not); as for ideal case $T_1 N_0$ (with previous lumpectomy) this rate is 74%.

Thus it is now established that electron beams used for additional doses (as one-third of the total dose) produce excellent late results regardless of whether radiation sequelae are reduced to a minimum, both in terms of number and importance.

As a conclusion, here are our indications for this type of radiation therapy. For *stage I* (T_1, T_2/N_0) we prefer initial lumpectomy followed by irradiation delivered according to curative doses previously described. For *stage II* (T_1, T_2/N_1) and *stage III* (provided tumor diameter does not exceed 8 cm) we perform exclusive radiation therapy according to doses indicated earlier. For fast growing cancers (where surgery is obviously not applicable) we perform, whatever the stage may be, exclusive radiation therapy according to the curative doses described, after two or three synchronizing chemotherapy courses.

Electron Beam Therapy of Advanced and Recurrent Breast Cancer

F. Chu

High energy electron beam therapy has made major contributions to the radiotherapeutic management of breast cancer. Electron beam therapy has proven to be very effective in local tumor control and in sparing underlying normal tissues. When irradiation is applied skillfully, excellent results can be achieved.

At Memorial Sloan-Kettering Cancer Center we began electron beam therapy about 25 years ago, when a 24 MeV Allis-Chalmers betatron was installed in 1953. During the subsequent years many thousands of patients have been treated, the largest group being patients with breast cancer. In our previous publications [1-7], we have defined the indications for electron irradiation, described methods of treatment, as well as analyzed the results of treatment. The methodology for electron beam treatment planning, including provision for inhomogeneities, was also developed [8-12].

Treatment of Locally Advanced Breast Cancer

The cases which are most suitable for electron beam therapy in the range of 6 to 24 MeV are relatively small breasts, flat type of breasts, and supplementary (boost) treatment. For the treatment of relatively small breasts, we use a five-field technique [3, 7]. The breast is treated with parallel opposing tangential fields. The axilla, the internal mammary, and supraclavicular areas are treated with separate portals.

With the highest energy available to us at 25 MeV, large breasts with interfield separation of more than 16 cm are not suitable for electron beam treatment due to the fall-off of the dose at the center of the breast. These cases are usually treated with cobalt or megavoltage X-rays.

Electron beam therapy is useful in treating flat type breasts with a single anterior field. The set up is simple and easy, as compared to the tangential technique. However, caution must be exercised to avoid the use of unnecessarily high energies, which would result in undesired irradiation of the lung. We advise that all patients be observed closely during the course of therapy and when the tumor size reduces, the electron energy should also be reduced accordingly, in order to avoid or minimize the possibility of radiation induced pneumonitis.

A very popular use of electron beam is to give supplementary (boost) treatment. For example, after the breast has received 5000 rads in 5 weeks using megavoltage X-ray or cobalt therapy, the residual tumor mass may be given an additional dose of 1500-2000 rads in about 2 weeks, using electron beam.

Electron therapy is very effective in controlling locally advanced breast cancer. In a study to evaluate the results of electron irradiation of 122 patients with advanced cancer [2, 5], 60% achieved complete control, which means disappearance of the tumor. The mean duration of complete control was 16 months. Another 27% showed a partial control for a mean duration of 8 months. The total response rate was 87%.

Treatment of Local Recurrence

Local or regional recurrence of breast cancer is a common problem. There is no doubt in our experience that electron beam therapy is the best treatment for these lesions. Local recurrence of limited extent following mastectomy warrants aggressive treatment. Indeed, some patients may remain free of disease for many years after such treatment. The radiation field should cover the site of recurrence with a wide margin, in order to include other occult lesions which may exist in the neighboring skin. A tumor dose of 4500-5000 rads is given over a period of 4 1/2-5 weeks with a supplementary dose of 500-1000 rads in 1 week, given to any residual nodule or nodules.

Most patients with recurrence, however, have extensive local involvement or evidence of distant metastases. Only palliative therapy can be offered to these cases. The radiation dose used for palliation is about 3500-4000 rads, delivered in a period of 3-4 weeks.

All of our treatment plans are individualized and tailored to suit the patient's particular needs. We formerly did routine transverse tomograms to determine the thickness of the chest wall and to outline the anterior surface of the lung. Our isodose distributions are corrected for the lower density of lung tissue according to the procedure previously described by Laughlin et al [8, 9, 11, 12].

More recently, we have been using computerized (CT) scans for the tomography in treatment planning. Our staff has determined the relationship between "delta number" and electron density, and this conversion is automatically carried out on a pixel by pixel basis [10]. The quality of imaging of CT scans is far superior than that of transverse tomograms; CT scans not only show clear images of the tumor and its surrounding normal structures, but also provide the densities of various tissues, facilitating precision treatment planning. Not too infrequently a recurrent nodule, particularly in the parasternal region, appears superficial clinically but actually penetrates deep into the chest, demonstrated only on the CT scan. Treatment plans using CT scans will be published elsewhere.

Various techniques of treating the chest wall or lymph node areas have been described in previous publications [2, 3, 5, 6]. A few important points need to be reemphasized. Of course, the easiest and simplest way of treating chest wall lesions is through one single anterior field, if one field can incorporate all lesions with adequate margins. Appropriate energy is chosen depending on the thickness of the chest wall and the thickness of the lesion or lesions. One should also fully utilize polystyrene absorbers, to improve the dose distribution and to bring the maximum dose close to the skin surface. Since the chest wall is usually thin after a mastectomy, we have found that 6-8 MeV electron energies are most useful for the treatment of local recurrences. The cutaneous and subcutaneous lesions usually receive a homogeneous dose within the 90% isodose line and the radiation dose drops off sharply to 10% within the first 2-3 cm of the anterior lung surface.

When cutaneous lesions involve a wide area that cannot be encompassed by a single field, they can be treated through two or more fields. There should be a separation of

0.5-1 cm between the fields, in order to avoid a "hot spot" due to scattering associated with the electron beam. We also advise that the location of the separation be shifted during treatment to avoid a "cold spot". This is usually done by changing the location of separation after every 1000-1500 rads.

Cutaneous lesions involving a curved surface may be treated with angulated fields. Extreme care must be taken to avoid overdosage at the interface of two converging fields. A "hot spot" may cause fractures of the ribs [6]. The use of polystyrene absorbers at the interface of the fields will solve most of the problems [2, 5].

In a study which evaluated the local response of lesions to electron beam therapy [2, 5], 630 cases of chest wall recurrences were treated. Most patients had extensive involvement and many had prior radiation therapy with X-rays or cobalt. The results showed 74% complete control of local disease for a mean duration of 13 months and 10% partial control for a mean duration of 5 months. The total response rate was 84%. In an evaluation of 297 lymph node areas treated by electron beam therapy [2, 5], 71% were controlled for a mean duration of 16 months, and 8% were controlled partially for a mean duration of 9 months. The total response rate was 79%.

Most patients tolerated electron beam therapy extremely well. There was moderate to brisk skin reactions, which usually subsided, with minimal or no late consequences. The incidence of radiation pneumonitis is very low, below 4%.

Recently we have reviewed and analyzed the results of treatment of 215 patients with limited local or regional recurrence to determine the survival status, as well as local disease control [4]. Most patients were treated with electron irradiation. These patients did not have evidence of distant metastases at the time of treatment.

Of the 215 cases treated, complete control of the local disease was achieved in 67% of cases for a mean duration of 32 months and a median duration of 22 months. Another 24% showed partial control for mean and median durations of 11 months and 8 months respectively. The total response was 91%. Most patients were free of local problems during this prolonged period of time.

The survival data showed that of the 215 patients evaluated, 44 (21%) survived 5 years, and 10 (5%) survived 10 years following radiation therapy of recurrent disease. There were seven patients, or 3% who were free of cancer at 5-15 years.

Another significant finding is that patients with complete control of the local disease had a higher survival rate (26%) than those with partial or no control (9% and 0% respectively). Intensive radiotherapy to patients with limited recurrence is therefore, definitely worthwhile.

Summary

High energy electron beam therapy is a valuable method of treating advanced and recurrent breast cancer. It offers versatile treatment techniques, excellent dose distributions, and effective local tumor control. Electron beam therapy occupies an important place in the radiotherapeutic management of breast cancer.

References

1 Chu FCH (1965) Experience with electron irradiation of breast cancer. In: Zuppinger A, Poretti G (eds) Symposium on high-energy electrons. Springer, Berlin Heidelberg New York

2 Chu FCH (1968) The role of electron beam therapy in the treatment of cancer. Front Radiat Ther Onc 2:224-237

3 Chu FCH (1971) Electron treatment of mammary cancer. Proceedings of the XII Postgraduate Course on Clinical Oncology. National Cancer Institute, Milan, Italy

4 Chu FCH, Lin FJ, Kim JH, Huh SH, Garmatis GJ (1976) Locally recurrent carcinoma of the breast. Results of radiation therapy. Cancer 37:2677-2690

5 Chu FCH, Nisce LZ, Baker AS, Sattar A, Laughlin JS (1967) Electron-beam therapy of cancer of the breast. Radiology 89:216-223

6 Chu FCH, Nisce LZ, Laughlin JS (1963) Treatment of breast cancer with high-energy electrons produced by 24-Mev betatron. Radiology 81:871-880

7 Chu FCH, Scheer AC, Gaspar-Landero J (1960) Electron-beam therapy in the management of carcinoma of the breast. Radiology 75:559-567

8 Dahler A, Baker AS, Laughlin JS (1969) Comprehensive electron-beam treatment planning. Ann NY Acad Sci 161:198-213

9 Holt JG, Mohan R, Caley R, Buffa A, Reid A, Simpson LD, Laughlin JS (1977) Memorial electron-beam AET treatment planning system in practical aspects of electron beam treatment planning. Med Phys Monogr 2:70-79

10 Hsi SC, Laughlin JS, Miller D, Masterson ME, Simpson L, Pentlow K (1978) An experimentally based computation approach for the relationship between CT number and electron density for radiation therapy treatment planning (works in progress). Presented at the Sixty-fourth Annual Meeting of the Radiological Society of North America, Chicago, Illinois

11 Laughlin JS (1965) High-energy electron treatment planning for inhomogeneities. Br J Radiol 38:143

12 Laughlin JS, Lundi A, Phillips R, Chu F, Sattar A (1965) Electron-beam treatment planning in inhomogeneous tissue. Radiology 85:524-530

Postoperative Treatment of Breast Cancer

J.P. Paunier

The Radiotherapy Center of the Cantonal Hospital in Geneva, which was inaugurated in January 1968, was originally equipped with a Siemens 18 MeV betatron in addition to a cobalt unit. In 1974 the betatron was replaced by a Varian 18 MeV linear accelerator because the output was too low for photon treatment. We had used the betatron exclusively for electron treatments. Our protocols for breast cancer treatments at the Cantonal Hospital in Geneva have been adapted from Fletcher's techniques applied at the M.D. Anderson Hospital, Houston, Texas (Fletcher 1972). However, many of our patients did not have a Halstead mastectomy. They had either a Patey operation or a simple mastectomy, with or without axillary sampling. These differences in surgical procedures can be easily explained by the sources of patient referrals. Our patients came not only from University Clinics, where the treatments are standardized, but also from practicing physicians in the city who perform various surgical techniques.

During the 6 years between 1968 and 1973, we gave postoperative therapy to 155 patients who had a Halstead operation, 107 who had a Patey operation, and 50 who had undergone a simple mastectomy. Of the total 313 patients treated by surgery followed by radiotherapy, 92 died either of their cancer or of other causes, giving a 5-year survival rate of 70.6%. We do not yet have a 10-year follow up. In evaluating the results of treatment of breast cancer a 10-year survival rate would be much more meaningul than a 5-year rate. Our results, however, compare favorably with the 5-year survival rate of 75% reported in the literature for patients with negative axillary nodes after a radical mastectomy. In our series there were patients with positive axillary nodes and patients with a central or medial tumor with negative axillary nodes. Montague (1978) and Fletcher (1972) have reviewed the survival of patients who were treated by a radical mastectomy, with or without radiotherapy. There were 287 patients who had a favorable prognosis and did not receive postoperative treatment. Of these, 71% were alive at 5 years and 54% at 10 years. There were 356 patients whose prognosis was less favorable due to more extensive involvement of the axillary nodes or to the location of their tumor in the central or medial quadrant of the breast, who received postoperative irradiation. The results have shown a 5-year survival rate of 71%, and a 10-year rate of 56%. As to preoperative irradiation, it was used on 438 patients with unfavorable prognosis (palpable nodes, inflammatory breast after biopsy, etc.), and resulted in 71% and 57% survival rates for 5 and 10 years, respectively. This study not only shows that radiotherapy does not have an unfavorable effect on the patients' survival, as some authors claim, but also indicates that it can prolong the survival of some patients with positive axillary nodes, or with a tumor located centrally or medially.

Between 1968 and 1973, we treated 133 post radical mastectomy patients by using the so-called peripheral technique (supraclavicular and internal mammary irradiation).

Of these, 75 patients had positive axillary nodes and 58 had either a medial or a central tumor and negative axillary nodes. This group had an 81% survival at 5 years. These good results were due mainly to the fact that the original tumor was less than 5 cm in diameter, and that the positive nodes in the axilla were never more than five. All patients with more than five positive axillary nodes and/or with a tumor more than 5 cm in diameter were given a comprehensive radiotherapy which not only included all the nodal areas but also the chest wall. At that time, telecobalt, not electron beam therapy, was used for comprehensive treatments.

Our irradiation technique for the peripheral lymphatics treatment is similar to the one used at the M.D. Anderson Hospital. We irradiated the supraclavicular area and the internal mammary chain, delivering a total given dose of 5000 rads in 4 weeks. The energy used was 12 MeV for the supraclavicular area and 15 MeV for the internal mammary chain.

For the 30 patients who did not have a thorough axillary dissection, we added a direct axillary field using 12-15 MeV according to the thickness of the tissue between the axilla and the supraclavicular field. Five of these patients developed complications of brachial plexus neuritis, which appeared between 2 and 5 years after radiotherapy. All five of them were then subjected to surgery which showed typical demyelination of the nerves. A careful dosimetric study was carried out and it was found that part of the brachial plexus had received a dose corresponding to an NSD of over 2000 rets.

A recent study [1] shows that for a 3-week treatment, a dose of 5000 rad produces a 50% probability of brachial plexus neuritis.

In all of our cases which showed complications we had, to some extent, exceeded that dose at one point of the irradiated volume or another. We believe that irradiations of the axillary area as well as of the homolateral supraclavicular area with perpendicular fields with electrons should be banned. Our experience has demonstrated that this technique could be dangerous in producing complications, and we suggest that high energy photons be used instead. As matter of fact, we believe that the electron beam is not quite stable enough and that we cannot rely totally on the theoretical calculations concerning depth dose at the juxtaposition of two fields which are perpendicular to each other. After the appearance of our first radiation neuritis we gave up this method, but the latent period of the neuritis lasted up to several years, four cases appearing after the first one.

We have modified our technique since acquiring the Varian 18 MeV accelerator. Patients treated according to the peripheral technique receive telecobalt radiations to the internal mammary chain and to the supraclavicular area. Patients who need comprehensive postoperative irradiation are treated by telecobalt to the nodal areas, and by the electron beam of 6 MeV to the thoracic wall. It should be noted that the beam localizer of the Varian 18 MeV for electron treatments is not as easy to use as the one of the Siemens 18 MeV betatron.

None of the patients irradiated have had either supraclavicular or internal mammary chain recurrences. Although the number of patients treated was limited (133), we believe the data do indicate that a "given dose" of 5000 rads in 4 weeks is sufficient to control possible subclinical involvements. In order to avoid significant skin reactions, we have reduced the given dose to 4500 rads, which has not decreased the efficiency of the treatment in terms of local control.

Conclusions

The results of postoperative treatment of 133 patients using the peripheral technique [2] have shown an excellent local control and a satisfactory 5-year survival rate. It should be pointed out, however, that these selected patients had a relatively good prognosis. It should be emphasized that the use of perpendicular fields with electron beams must be avoided because of the high probability of nonhomogeneity of the dose and the possible instability of the beam.

References

1 Cohen L, Svensson H (1978) Cell population and dose-time relationship for post irradiation injury of the brachial plexus in man. Acta Radiol Oncol Radiat Phys Biol 17:161-166
2 Fletcher G (1972) Text book of radiotherapy, 2nd edn. Lea & Febiger, Philadelphia
3 Montague E (1978) Refresher Course of the "Radiological Society of North America". Chicago

Discussion of Breast Cancer

A. Zuppinger

There is little doubt that by electrons alone or in combination with photons the local control of breast cancer is possible with a high probability of cure. In many cases this treatment enables conservative surgery or even radiotherapy alone. The patients who do not survive die mainly due to distant metastases. It is generally assumed that these metastases are due to the natural tendency of breast cancer to metastasize. The other logical possibility that these metastases may be, at least partially, secondary to medical or surgical manipulations has not been examined up to now. It is possible to avoid this important question by irradiating such tumors before any surgical procedure has been done. Probably relatively low irradiation doses damage mainly the fast growing peripheral parts of the tumor, so that embolism does not take place or with a reduced percentage. In 1959 we began in patients with clinical and radiologic findings very suspicious of breast cancer to irradiate before a biopsy had been performed (prebioptic irradiation). We applied in tangential fields, with high speed electrons of 30-35 MeV, 2500-3000 rads in 10 to 14 days. Two to 5 days later surgical biopsy was performed and in those patients with positive findings, radical mastectomy or in a few patients because of advanced age, simple mastectomy, was carried out immediately afterwards. To allow a comparison to patients with operation and postoperative irradiation, we included postoperative irradiation in the prebioptic patients, so that both groups recived the same total dose. In 31 T_1N_0 and T_2N_0 patients with preoperative irradiation no distant metastases appeared and no local or regional metastases were observed in 5-10 years observation time. The postoperative group showed 23% in the 5-year period and 32% distant metastases in the total observation time. The cure rate of 68%

in the postoperative group rose to 93% in the prebioptic group. In the T_2N_1 group results have been even better. The cure rate rose from 40% to 78% in the 5-year period. These differences are statistically significant.

It is very probable that we produce a relatively high percentage of distant metastases by surgical biopsy or the operation itself. It has been very difficult to convince the physicians and surgeons to send their patients to prebioptic irradiation. Thus, the number of treated cases is relatively limited. We would like to recommend a clinical experiment on a broad base.

The prebioptic irradiation is certainly also possible with ^{60}Co or high energy photons. Electrons are, most probably, more effective because of their physical properties.

The original paper with more details has been published in the meantime in the book: Krokowski E (1979) Neue Aspekte der Krebsbekämpfung. Die präbioptische Bestrahlung beim Mammanarzinern. pp 101-107. Stuttgart, Thieme.

Carcinoma of the Bladder Treated with High Energy (33 MeV) Electrons

C. Botstein and S. Kalnicki

Introduction

The treatment of advanced bladder carcinoma is characterized by high morbidity and low survival rates. More than half of the deaths are due to persistent or recurrent local disease. Therefore, current treatment modalities must emphasize local control with minimal morbidity.

The purpose of this study is to review the results of treatment by primary electron beam irradiation at the Radiotherapy Department at Montefiore Hospital and Medical Center.

Materials and Methods

A total of 144 patients received primary electron beam treatment for carcinoma of the bladder at Montefiore Hospital from 1961-1976. There were 101 male and 43 female patients, with ages ranging from 55 to 90 years (peak 70-75 years).

Of these patients 117 (81%) had stages B2 to D (or advanced disease), according to the Jewett-Marshall classification.

Treatment was given utilizing the 33-MeV electron beam generated by the Brown-Boveri betatron (Askleptikon). A single anterior pelvic fied was used. During the early years of treatment, an anterior rectangular portal measuring 8 x 10 cm was utilized. More recently, the field was enlarged to 12 x 2 — 14 x 4 cm, in the diamond position to encompass the entire true pelvis.

In preparation, the bladder was initially localized with iodine contrast media and the anterior diamond pelvic field outlined. A large Foley catheter was then inserted into the rectum and its balloon filled with 20-30 cc iodine contrast. A lateral X-ray was taken, and the point of minimum dose was determined at the anterior surface of the rectal balloon. All bladder dosages described were measured at this point (Fig. 1).

The rectal balloon characteristically elevated the floor of the bladder 1-3 cm, increasing the dose by 10%-20%, and displaced the posterior rectal wall which received less than 20% of the given dose.

Dosages ranged between 6000 and 7000 rad, given in 50-60 days, usually on a three-times-a-week fractionation schedule.

Results

Of the 144 patients, 116 were evaluable for the study. All those who did not receive a full course of treatment were excluded, together with those who died immediately

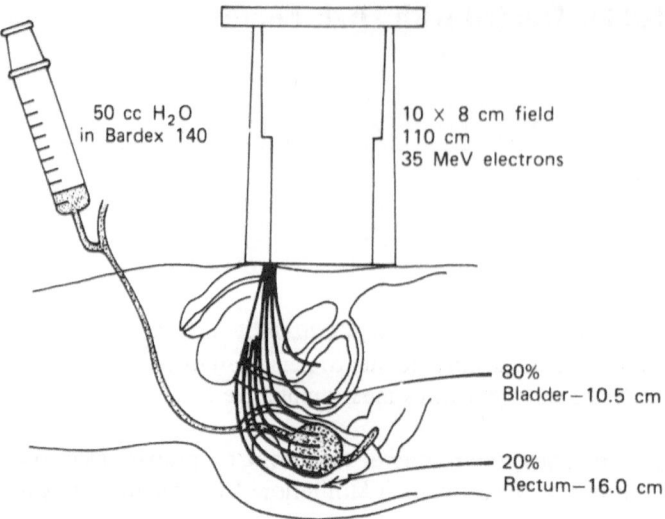

Fig. 1. Patient set-up for treatment of carcinoma of the bladder with 33 MeV electron beam

after the treatment, either because of advance disease or associated condition. Tables 1 and 2 show the overall and the sex-related survival, respectively. The overall 5-year survival was 14%, 8.4% for males and 17% for females. The better survival among women resulted from the supplementary use of intravaginal and/or intrauterine applicators. In the early years, the radium Fletcher after-loading system was utilized; later cesium-137 microsources were used with the Botstein-Zacharopoulos after-loading system.

Table 1. Òverall survival of patients. Carcinoma of bladder stages B_2-D

Survival in years	No. of patients	%
1	47/116	41
2	41/113	36
5	16/113	14

Table 2. Sex-related survival. Carcinoma of bladder stages B_2-D

	Male		Female	
Survival in years	No. of patients	%	No. of patients	%
1	27/83	32	20/33	61
2	23/83	28	16/30	53
5	7/83	8.4	5/30	17

To evaluate the efficacy of the electron beam, one has to compare this group of patients with another, treated primarily with photons. The effective range of the electron beam is limited to a depth which is approximately one-third of the maximum energy of the beam. Our treatment policy required that the tumor be encompassed by at least the 65% isodose curve. Thus, patients with a large sagittal diameter were unsuitable for treatment with electrons. They were treated primarily with high energy photons, electrons being occasionally utilized for supplementary irradiation. It should be emphasized that it was the size of the patients and not the characteristics of the tumor that determined the treatment modality. Although survival at 1 year was better for the photon group, survival at years and beyond was better for the electron group (Table 3).

Table 3. Comparison between electron beam and photons as primary mode of treatment for advanced bladder carcinoma

Survival years	% Survival	
	Electrons (%)	Photons (%)
1	41	50
2	36	9
5	14	0

The low incidence of side-effects and complications was one of the greatest advantages of the electron beam. The majority of patients presented with mild dysuria and dry desquamation at some point during the course of treatment, while some evidenced mild moist desquamation. The absence of rectal symptoms was striking. Late complications were proctitis requiring colostomy (in the female patient treated with combination of electron beam and intracavitary irradiation), contracted bladder (one patient), and enteritis (narrowing of the ileum in the gastrointestinal series in one patient). The only lethal complication was hemorrhagic cystitis and urosepsis in one patient retreated with cancericidal dose 6 years after primary irradiation.

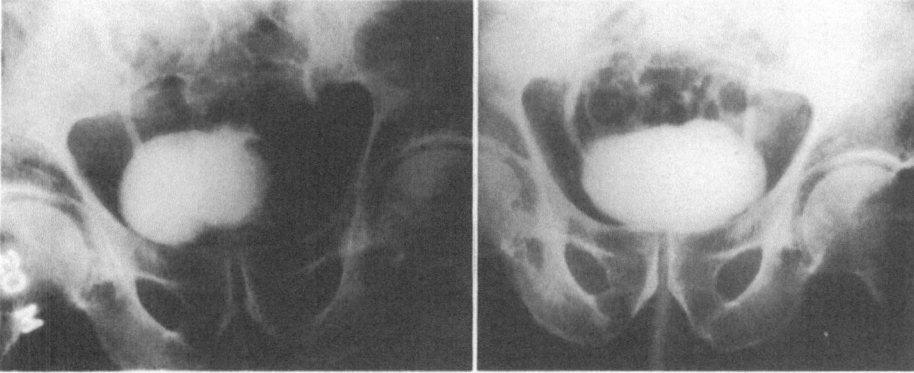

Fig. 2. Intravenous pyelogram before and after electron beam treatment

Discussion

The finite depth of penetration of the electron beam makes it theoretically ideal for treating anterior and sparing posterior structures. This permitted the administration of bladder doses that were clearly beyond rectal tolerance.

The patients referred for radiation therapy at Montefiore Hospital and Medical Center were inoperable due to advanced stage or associated medical problems. The average age of our patients was 10 years above that in most surgical series, and yet our figures were comparable. Whitmore and Marshall obtained a 15% 5-year survival with radical cystectomy for stages B2 and C, together with 51% morbidity. Our series, with an overall 5-year survival of 14% and a 2.7% morbidity, was based on patients of advanced age, failures of previous treatment, and late stage lesion.

In conclusion, high energy electrons are an effective and well-tolerated modality for treatment of carcinoma of the bladder.

Clinical Applications of High Energy Electron Beams: the Pancreas, Pleura, and Spine*

R.R. Dobelbower, Jr., K.A. Strubler, and I. Vaisman

Introduction

The role of radiation therapy in the treatment of deeply situated lesions, as well as those more superficially located, has been restricted, in part, by closely associated radioresponsive normal tissues. The control of neoplastic disease by radiation depends on the delivery of tumoricidal doses to the neoplasm while sparing adjacent critical structures. With the advent of high energy electron beams, dose distributions can be obtained which enhance the clinician's ability to achieve this goal.

Electron beams have now gained wide clinical acceptance by virtue of their sharp and rapid fall-off of dose with depth. In recent years, electrons have been commonly used for the treatment of nodal disease, chest wall irradiation for breast carcinoma, a variety of skin lesions and other somewhat superficial areas where electrons are needed to boost the primary photon beam radiotherapy. This paper briefly describes our experience with the application of high energy electron beams to the pancreas, pleura, and spine.

Distributions utilizing high energy electrons as well as photons were obtained from a 45 MeV Brown-Boveri betatron. This accelerator, in use in our department for the last 5 years, is capable of generating clinically useful beams of 45 MeV (peak) photons and 5-45 MeV electrons. Our institution sees a total of about 1000 new cancer cases annually, and we have routinely employed electron beam therapy in the management of a variety of clinical problems (Table 1). In addition, three categories of neoplasm, those of the pancreas, pleura, and spine, have been noted to pose special problems that seem best managed with the help of high energy electron beams.

Pancreas

Cancer of the pancreas is the fifth leading cause of all cancer deaths in the United States, and there has been no appreciable change in the prognosis in this disease over the last three decades. The incidence of pancreatic carcinoma is increasing [3, 8]. Only about 10% of patients are candidates for radical surgery, and the surgical mortality exceeds the 5-year survival rate [2]. The overall 5-year survival is only about 2% [6, 8]. Pancreatic cancer often kills by local growth without widespread metastasis [4]. It is estimated that 40%-60% of patients have advanced nonresectable disease at diagnosis without evidence of distant metastasis. If this group of patients could be treated effectively to achieve local control, the overall survival should be improved.

* Supported in part by U.S. Public Health Service Grants CA-09137 and CA-11602 from the National Cancer Institute, U.S. Department of Health, Education and Welfare and by a grant (JCFC 26313) from the American Cancer Society

Table 1. Electron beam sites of clinical use

1) Plevis
    Bladder
    Sacral hollow
2) Perineal region
    Anus
    Perineal body
    Vulva
3) Abdomen
    Abdominal wall
    Bile duct
    Gall bladder
    Pancreas
4) Breast
    Boost to primary lesion
    Chest wall
    Mediastinum
5) Head and neck
    Boost to various primary lesions
    Salivary glands
    Thyroid
6) Nodes
    Axillary
    Cervical
    Inguinal
    Internal mammary
7) Other
    Penis
    Pleura
    Skin
    Spine

Thirty-six patients with biopsy-proven, unresectable adenocarcinoma of the pancreas were treated with precision high dose (PHD) radiotherapy at our institution. No patient with hepatic or distant metastasis was treated, but all patients had disease extending beyond the pancreas. Patients ranged in age from 35 to 83 years, with a median age of 65 years. The most common symptoms were jaundice, pain, and weight loss. Radio-opaque clips were placed at the gross tumor margins at laparotomy in all but one patient. The following studies were also used to determine, as precisely as possible, not only the location of the tumor but the adjacent normal structures as well: upper gastrointestinal radiography, intravenous pyelography, abdominal ultrasound, and computerized tomography [4].

External beam irradiation was delivered from our betatron with 1-3 cm margins around the tumor. If the posterior margin of the target volume was 12 cm or less from the anterior surface of the abdomen (19 patients), a "mixed beam" technique was employed with opposed lateral 45 MeV (peak) photon fields and an anterior mixed beam using 15-35 MeV electrons and the 45 MeV (peak) X-rays (Fig. 1) [6]. The choice of electron energy was based on the posterior extent of the target volume and was de-

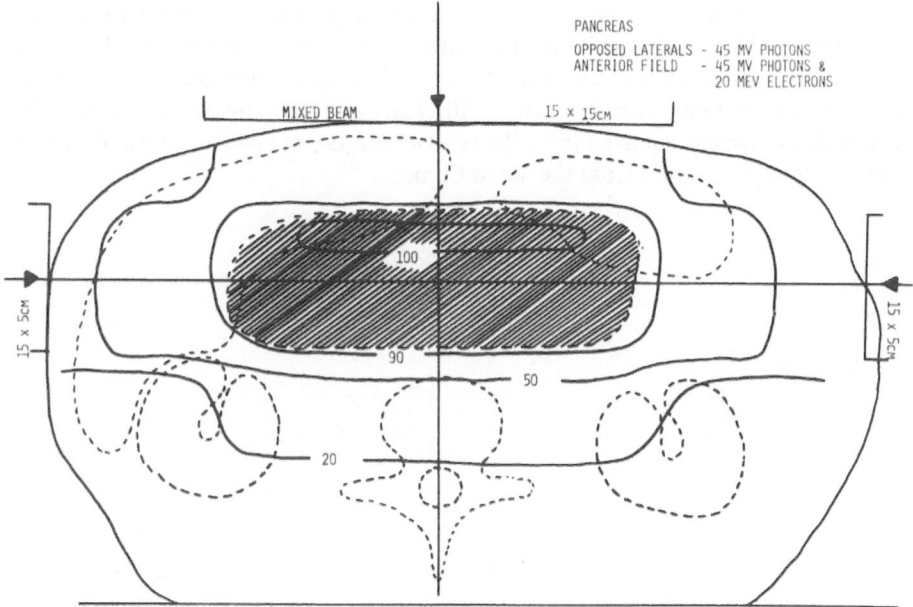

Fig. 1. Isodose distribution. Opposed lateral 45 MeV (peak) photon beams and anterior mixed beam (50%, 45 MeV [peak] photons, 50%, 20 MeV electrons)

signed to minimize kidney and cord dose. If the posterior margin of the target volume was more than 12 cm from the anterior surface of the abdomen (17 patients), mixed-beam dose distributions appeared inferior to pure photon beam distributions, and patients were treated with 3-field or 4-field photon beam techniques [4].

Where indicated, beams were extensively shaped with blocks or castings of low melting temperature alloy to conform the isodose distribution to the tumor profile. Tumor minimum doses (90% or 95% isodose line) of 180 rad (centigrays) were delivered 5 days per week to total doses of 5900-7000 rad (centigrays) over 7-9 weeks. Lesions situated laterally in the tail of the pancreas can also be treated by a single lateral beam of high energy electrons or a single lateral mixed beam.

No patient has been lost to follow-up, which ranges from 3-67 months post diagnosis. Actuarial survival data was determined by the Berkson-Gage method [1]. Twelve patients received adjuvant chemotherapy at the discretion of referring physicians [4, 5]. Four patients failed to complete treatment as planned. Treatment was interrupted for two other patients because of intercurrent illness. Six patients reported significant nausea, diarrhea, or anorexia during the course of treatment. Of 27 patients at risk for 6 months or more, seven developed clinically significant delayed reactions (gastritis or gastrointestinal bleeding); however, in five of the seven patients reactions coincided with local recurrence of cancer. Mild pancreatic insufficiency was observed in seven patients, and two patients developed elevated fasting blood sugar levels after PHD radiotherapy. No patient has died of radiation complications, and clinically significant hepatic or renal or spinal cord damage has not been observed.

Appetite improved in 6 of 12 anorexic patients, and only 15 patients lost 5 lb or more during therapy. The weights of 21 patients either increased or remained stable. Relief of pain was observed in 20 of 29 patients. Figure 2 compares actuarial survival of patients with unresectable disease treated with PHD radiotherapy to that of patients with resectable disease reported from the Massachusetts General Hospital and that of untreated patients reported from the Mayo Clinic.

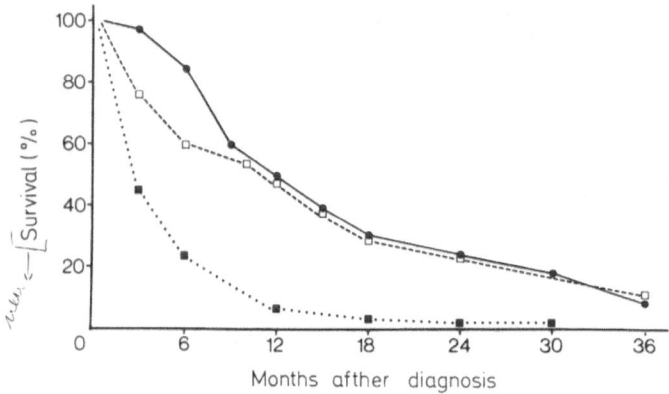

Fig. 2. Comparison of survival rates of: —●— 36 patients treated with PHD radiotherapy (Thomas Jefferson University Hospital); - - -□- - - 31 patients treated with radical surgery (Massachusetts General Hospital); · · ·■· · · 145 patients receiving no treatment (Mayo Clinic)

The apparent initial advantage of radiotherapy over surgery may reflect acute surgical mortality; otherwise, the two survival curves are quite comparable to the 3-year point. This favorable comparison indicates that this therapeutic modality has the potential for delivering curative doses to reasonably localized, unresectable pancreatic cancer while sparing adjacent structures such as the spinal cord, kidney, liver, and gut.

Pleura

Malignant pleural mesothelioma is a rare but highly lethal neoplasm arising from the mesothelial lining of the thoracic cavity. This disease has long been regarded as "radioresistant." This is probably due, in part, to the circumstance that the pleura encases the lung, an organ of rather low radiotolerance. It has not been established that mesothelioma is inherently nonresponsive to radiation. In fact, one group of investigators has shown enhancement of survival with increase in radiation dose above 3500 rads (centigrays) using conventional treatment techniques [7, 9]. With the availability of photon and electrons of various energies, we may now be able to deliver a high dose to pleural tumors and still spare the underlying normal pulmonary tissue.

Nine patients ranging in age from 46 to 75 years (median age: 61 years) have been treated with combined photon/electron techniques for biopsy-proven malignant mesothelioma. Presenting signs and symptoms included dyspnea at rest, exertional dyspnea, chest pain, hemoptysis, pleural effusion, chronic cough, and sensation of fullness in the chest. Nonspecific symptoms of weight loss, fever, weakness, and chills were also present. A past history of asbestos exposure was elicited from eight patients.

The involved areas of the lung and the mediastinum were treated with shaped, parallel opposed beams of 4 MeV (peak) photons. The central normal pulmonary parenchyma was fully blocked, both anteriorly and posteriorly. These blocked central areas were concurrently treated with 10-20 MeV electrons. The energy was selected in accordance with the chest wall thickness as measured by ultrasound.

Using the combined shaped electron and photon beams, 2000 rad (centigrays) were delivered to the pleura in 2 weeks in ten equal increments. All fields were treated daily. This was followed by a 2-week rest and the cycle was then repeated. After delivery of 4000 rad (centigrays) in 6 weeks, the photon fields were modified to avoid the spinal cord and, after a second 2-week rest, the final 2000 rad (centigrays) was delivered in 2 weeks for a total tumor dose of 6000 rad (centigrays) in 10 weeks in a double split course fashion. Figure 3 radiographically demonstrates that normal lung tissue can be spared with this technique and shows a typical photon portal film.

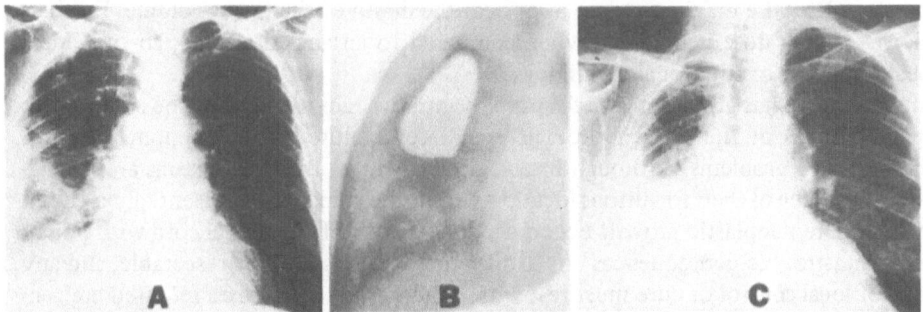

Fig. 3 A-C. Radiographs demonstrating that the normal pulmonary parenchyma can be spared by the mesothelioma technique. **A** Chest radiograph prior to treatment. **B** Treatment portal film. **C** Chest radiograph one-year post-treatment

Two patients died of cardiorespiratory failure during the first of the three treatment cycles. In the seven patients who completed treatment, pain and cough were generally relieved, and dyspnea and unspecific symptoms were improved in most. There was only slight improvement of pulmonary function. One patient was treated with 3000 rad (centigrays) in 3 weeks, a 2-week rest, and an additional 3000 rad (centigrays) in 3 weeks for a total dose of 6000 rad (centigrays) in 8 weeks. He developed symptomatic radiation pneumonitis. Two patients are alive at 20 and 40 months post diagnosis. One of these survivors is a patient with radiation pneumonitis and is apparently free of disease. The other survivor has required palliative irradiation for presumed osseous metastasis.

This technique does seem to have significant palliative value, but it remains to be seen whether the time and effort required for simulation, block making, treatment planning, ultrasound, imaging, patient set-up, etc., is justified in terms of local control of mesothelioma.

Spine

Vertebral metastasis are exceedingly common, especially from primary malignant tumors of the breast, prostate, and lung. In such cases local radiotherapy has been repeatedly shown to have significant palliative value. The morbidity of such treatment is related to the adjacent normal tissues such as pharyngeal and esophageal mucosa, gut, kidneys, spinal cord, and skin. High energy electrons "mixed" with photons provide simplicity of treatment as well as superior dose distributions when compared to conventional single posterior photon beams.

Photon and electron beams are weighted one-to-one since varying the electron beam energy has a similar effect to differential weighting. The summated distribution offers several advantages. In the region of the lumbar spine, a single posterior mixed beam (50%, 45 MeV [peak] photons and 50%, electrons of appropriate energy) provides a lower subcutaneous dose than a single posterior beam of Cobalt-60 or 4 MeV [peak] radiation and a lower gut dose than a single posterior high energy photon beam and also places the maximum dose at or near the depth of the target volume. Further, a lower kidney dose is obtained when compared to an angled wedge pair of 4 MeV [peak] photons or rotational techniques.

This technique has also been used for primary curative radiotherapy of the spinal cord. Ependymomas of the cervical-thoracic spinal cord, although uncommon, present a unique set of problems. Although histologically benign, these neoplasms are "malignant" by virtue of their fortuitous location within the spinal canal, essentially a closed space, where neoplastic growth occurs at the expense of the spinal cord with potentially catastrophic consequences. As a rule, these lesions are not resectable, and any hope of local control or cure must rest with radiotherapy. The doses required are generally considerably higher than those needed for palliation of vertebral metastasis. Hence, avoiding adjacent normal structures becomes even more important. The simplicity and effectiveness of the mixed beam technique has obviated possible problems associated with other techniques in the treatment of this disease.

Summary

The clinical use of high energy electron beams appears justified from a dose distribution point of view. In addition, there has been an important emphasis in recent years in devising new methods of localizing tumor masses and critical normal structures. It is somewhat disturbing to see illustrations of various techniques which nicely localize critical structures within the body only to find that the treatment technique used was a pair of simple, parallel opposed fields. In many situations, the high energy electron beams, often used in combination with high energy photons, can be used with nearly equal simplicity while, at the same time, obtaining a better conformation of the high dose volume to the tumor. As new methods for localizing tumor masses and normal tissues are developed, innovative approaches for dose delivery must be utilized. We believe that with the additional capability of high energy electron beams, it may now be possible to provide improved palliation if not definitive radiotherapeutic management of localized disease for a number of both superficial and deeply situated neoplasms.

References

1 Berkson J, Gage RP (1950) Calculation of survival rates for cancer. Prof Mayo Clin 25:270-86
2 Borgelt BB, Dobelbower RR, Jr, Strubler KA (1978) Betatron therapy for unresectable pancreatic cancer: A preliminary report. Am J Surg 135:76-80
3 Cancer Statistics (1977) CA − Cancer J Clin 27:26-41
4 Dobelbower RR, Jr (1979) Cancer of the pancreas − Radiation therapy. Advances in medical oncology, research and education, vol 1, 9. Digestive cancer. Thacher N (ed). Pergamon Press, Oxford
5 Dobelbower RR, Jr, Borgelt BB, Suntharalingam N, Strubler KA (1978) Pancreatic carcinoma treated with high-dose, small-volume irradiation. Cancer 41:1087-92
6 Dobelbower RR, Jr, Strubler KA, Suntharalingam N (1976) Treatment of cancer of the pancreas with high-energy photons and electrons. Int J Radiat Oncol 1:141-146
7 Eschwege F, Schlienger M (1973) La Radiotherapie des mesotheliomas pleuraux malins. A propos de 14 cas irradies a doeses elevees. J Radiol Electrol 54:255-259
8 Levin DL, Connelly RR (1972) Cancer of the pancreas. Available epidemiologic information and its implications. Cancer 31:1231-36
9 Schlienger M, Eschwege F, Blache R, de Pierre R (1969) Mesotheliomas pleuraux malins. Etude de 39 cas dont 25 autopsies. Bull Cancer 56:265-308

Electron Therapy in Tumors of the Digestive Tract

A. Zuppinger

Epithelial gastrointestinal tumors are generally considered to be radioresistant. Surgery is the dominant therapy but the final and global results are not at all satisfactory. Many attempts have been made to raise the cure rate. Postoperative irradiation has, up to date, not reduced the number of local and regional recurrences. It has, in addition, the drawback of being poorly tolerated. Also the surgeons could not be convinced to use preoperative irradiation on a broad scale, though some clinical trials have shown that in patients with rectal and sigmoid carcinomas and lymph node involvement the results are better than in the control group.

Many, if not the majority, of the surgeons are reluctant to operate in previously irradiated areas. This is very difficult to overcome, even for patients with low irradiation doses, where no higher complication rate exists. Every new trial with a combination of surgery with radiotherapy has to face this psychological situation and to prove first the effectivness of ionization radiation on inoperable patients.

We conducted our first trials with electrons in 1958. A surgeon, Dr. Renfer, a good friend of mine, had operated on a 66-year-old woman (Fig. 1) because of an adenocarcinoma of the ascending colon. Half a year later she developed a tumor of the size of a child's head in the operation scar in the abdominal wall. The tumor was fixed to the deeper structures and had to be declared as inoperable. We made a crossfield irradiation in two series with 7300 rads in 64 days and we were highly surprised to see that the tumor had practically disappeared by the end of the treatment. Eleven months later the patient was operated on by the same surgeon because of a duodenal ulcer. No complications occurred either during the operation or in the postoperative phase. It may be that the ulcer arose in some connection with the irradiation.

No tumor tissue could be found histologically. The patient showed afterwards a localized induration in the abdominal wall which did not disturb her. She remained free of disease until she died of old age 14 years after our treatment.

A year later, the same surgeon showed me another patient operated on because of a mucus-secreting adenocarcinoma of the sigma, who developed a metastasis in the abdominal wall. He had excised it twice, but the third recurrence proved to be inoperable, so that the surgeon performed only a biopsy which gave the same histology. This histological type has been known since the time of Regaud to be practically radioresistant. I personally had never succeeded even with 30 MeV photons in really influencing such a tumor. I simply made a trial and applied a dose of 6040 rads in 38 days with 25 MeV electrons with a crossfire arrangement to protect the intestines. The tumor disappeared entirely, but the patient showed an induration at the site of the irradiation which did not disturb him. Seven years later another surgeon made a biopsy without informing us, but no tumor tissue could be found histologically. Some months later the patient died because of cerebral arteriosclerosis.

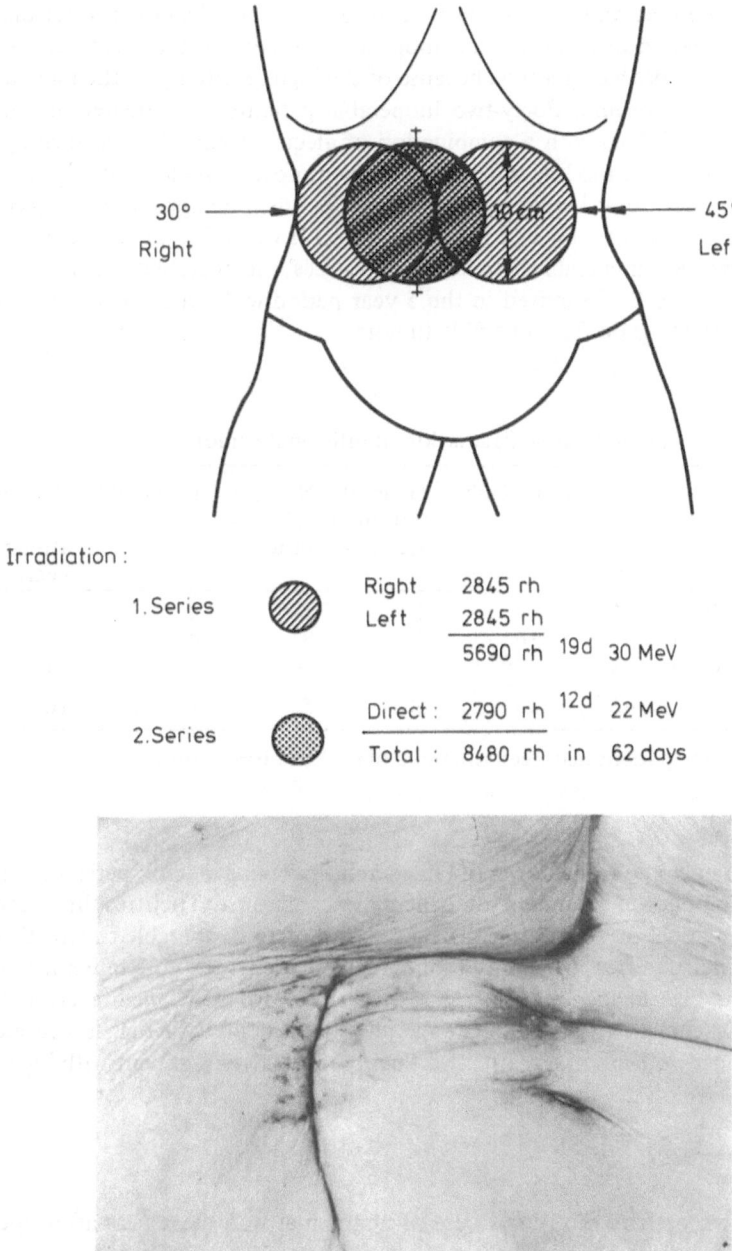

Fig. 1. 63-year-old woman with inoperable recurrence after resection of an adeno-carcinoma of the ascending colon. *Above:* crossfire irradiation in two series with a total dose of 7300 rads 64 days 31 irradiations. *Below:* same patient 6 years later, free of disease. The scar is that from the duodenal ulcer operation

These observations have been decisive for the application of electrons as preoperative irradiation in order to make inoperable patients operable, since a curative application is not possible, as a rule, because of the high sensitivity of the mucous membrane and the peritoneum. Forty-two inoperable patients were treated in this way (Table 1), most of them with a combination of electrons and photons, mainly because 30 and 35 MeV are too low to apply a sufficient dose with electrons alone in deeply situated tumors. Doses between 3000 and 4000 rads were applied. We succeeded in achieving an operable tumor in 57% of the cases. If we add to the operable patients also those who were inoperable because of metastases, the operability figure rises to almost 70%. But only 23% survived in the 5-year period without evidence of tumor, which is surprisingly good for inoperable tumors.

Table 1. Preoperative irradiation in intestinal tumors

	Number	Operable	Operable but metastases	No radical operation	Inoperable	Free of disease after 5 years	Free of disease, intercurrent death
Stomach	17	9	2	2	4	2	—
Rectum	24	14	3	3	4	7	1
Caecum	2	2[a]	—	—	—	1	1
	43	25	5	5	8	10	2

[a] One patient had been operable before the irradiation

Cramer and Dobelbower of Philadelphia have had much experience in a combined electron-photon treatment of pancreatic carcinoma. Their results in inoperable tumors are equal or even better than those of surgery in operable pancreatic cancer. If we restrict this treatment only to inoperable patients the cure rate must be low because of the high incidence of metastases. All the above-mentioned facts seem to be encouraging enough to promote preoperative treatment with high speed electrons which is most probably more effective than preoperative treatment with high energy photons.

Summary

Two patients with local cure of abdominal wall metastasis after operation for colon tumors are described. Curative treatment of intestinal carcinomas is not possible as a rule because of the high sensitivity of the mucous membranes and the peritoneum. After preoperative irradiation in 42 patients with inoperable intestinal cancers 57% became operable. Ten patients (23% of all preoperative treatments) survived the 5-year period without evidence of tumor. A broad clinical trial with preoperative electron beam irradiation is recommended. It is very probable that this is more effective than preoperative treatment with photons alone.

Reference

Zuppinger A (1972) Magen, Dünndarm, Colon. In: Zuppinger A, Krokowski E (eds) Radiation therapy of malignant tumours. Springer, Berlin Heidelberg New York (Encyclopedia of Medical radiology, vol XIX/1, pp 543-634)

Soft Tissue Sarcomas Treated by Electron Beams

R. Denepoux, J. Touchard, J. Pigneux, P. Richaud, and C. Lagarde

Soft tissue sarcomas are nonepithelial tumors arising from tissue other than bone (except for the reticuloendothelial system, subtentacular tissue of viscera and organs) and from the peripheral and sympathetic nervous system [7]. All sarcomas developed in viscera, Kaposi's sarcoma, soft tissue metastases, and soft tissue tumors in children [5] are excluded from this study.

Patients

From January 1971 through December 1977, 99 patients were treated at the Foundation Bergonie Cancer Center. Forty-two of them received electron beam therapy either alone or associated with another treatment.

Figure 1 shows the distribution of the 42 patients by age and sex. There was only one patient under 20 years, since sarcomas of soft tissue in children were not included in this study. The median age was 53 years. The topographic, histological, and clinical aspects of these tumors were very multifaceted as generally described [9]. Tables 1 and 2 show the distribution by anatomic site and histology. Clinically, the first symptom was generally a swelling, sometimes with pain; neurologic or vascular compression were uncommon (Fig. 2).

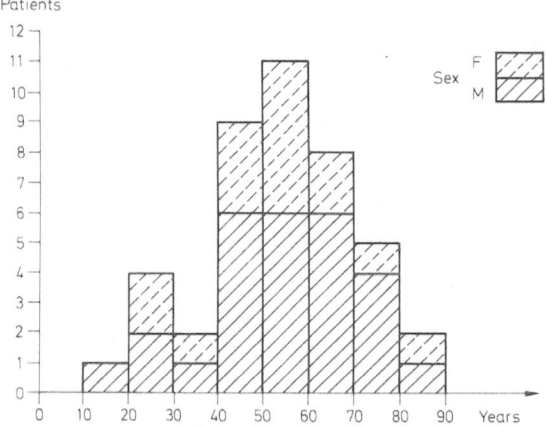

Fig. 1. Distribution of soft tissue sarcomas according to age and sex of patients

Table 1. Soft tissue sarcomas (42 patients): distribution by anatomic site

Chest wall	12
Pelvic girdle	10
Scapular girdle	6
Lower limb	5
Abdominal wall	5
Upper limb	3
Neck	1

Table 2. Soft tissue sarcomas (42 patients): distribution according to histology

Fibrosarcoma	14
Rhabdomyosarcoma	9
Schwannosarcoma	7
Alveolar sarcoma	4
Malignant mesothelioma	4
Malignant chordoma	2
Angiosarcoma	1
Liposarcoma	1

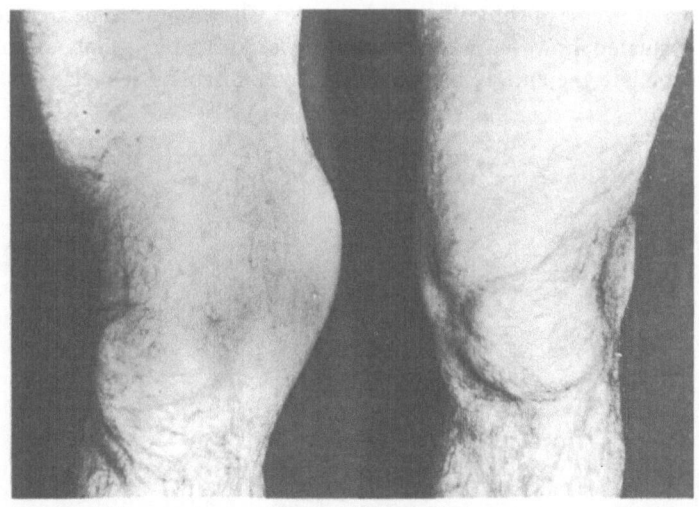

Fig. 2

There is no TNM classification for soft tissue tumors. We staged the tumors by size:

Stage I: maximum tumor axis less than 5 cm (7 patients)

Stage II: maximum tumor axis more than 5 cm
                or multifocal tumors   (25 patients)

Stage III: "historical" tumors   (3 patients)

Seven patients, referred to our institution after previous surgical treatment elsewhere (for which no details were available) were not staged. Five patients presented initially with metastatic lymph nodes and ten with distant metastases. The patients were first seen in consultation 3 months to 16 years after the onset of the first symptom. Twenty-five out of the 42 patients had received previous treatment before radiation therapy (14 patients had one previous treatment, four had two previous treatments, and seven had three or more previous treatments).

Methods

Electron beam therapy was performed for 27 patients in association with surgery, 4 times before and 23 times after the surgery. For the 15 patients treated by radiation therapy alone, 7 were treated by a combination of photons and electrons, and 8 by electron beam alone. Six patients were, in addition, submitted to a chemotherapy after radiation therapy.

The target volume was defined by clinical examination, radiography (including arteriography), xerography and more recently by computerized transverse tomography. The arteriography generally indicated either an abundant neovascularization, more important than the palpable tumor, or, more rarely, a nonvascularized mass (Figs. 3 and 4). We have to point out that very often thermography was useful in determining the local extent. Lymph node areas were treated only in patients with rhabdomyosarcoma or in case of clinically positive nodes. The volume to be treated was always largely evaluated; however, when the tumor was located in a limb, attention was taken not to irradiate the entire limb, so that delayed disabling sequelae would be avoided. The

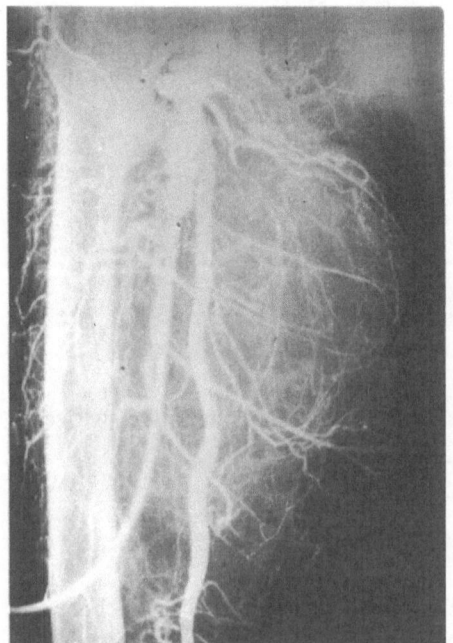
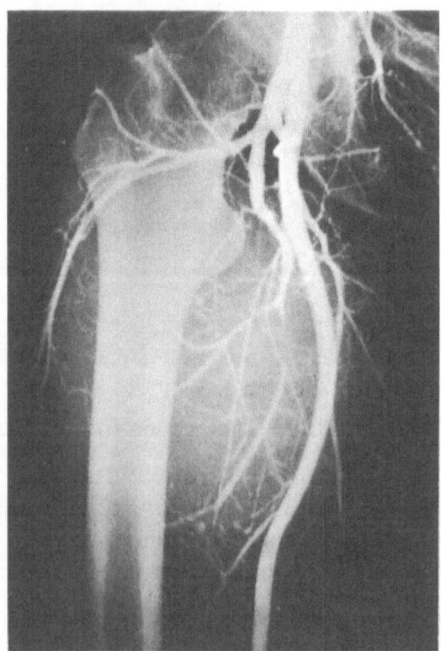

Fig. 3   Fig. 4

doses delivered were variable because the radiosensitivity of the tumors differ according to their histology.

Angiosarcomas were more radioresistant than lipo- or fibrosarcomas, but the doses delivered were in the range of 50-70 Gy, with a boost of 10-20 Gy given just on the strict tumoral volume or on the scar. For postoperative irradiation after total surgical excision, the dose was limited to 60 Gy. For some patients, the treatment was only palliative, antihemorrhagic or analgesic; in these cases the dose was only 30-50 Gy.

Results

Survival and local control are quite different in patients treated by surgery and radiotherapy and in patients not suitable for surgery and treated by radiation therapy alone. For the first group of 27 patients treated by combined treatment:

Seventeen are alive and disease free with a follow up period of 9 to 60 months.
Three are alive with local recurrence which appeared 2, 8, and 20 months after the completion of treatment.
Seven deaths occurred: three were due to local relapse at 12, 13, and 24 months, three to distant metastasis at 6, 15, and 36 months, and one to intercurrent disease.

In the second group of patients treated by irradiation alone, only 2 patients out of 15 are alive with a follow up period of 12 and 16 months. However, we have to point out that all these patients had very advanced disease and were in very poor general condition.

Tolerance. Since only one portal of the apparatus could be used, acute reactions were always localized. We generally observed moist reactions which healed quickly. Only one patient presented a delay in surgical scar healing. The incidence of severe subcutaneous fibrosis in patients treated with electron beams alone was low: one patient with a soft tissue sarcoma of the supraclavicular area who was treated by surgery and electron beams developed a late severe fibrosis with ulnar paresthesia. We did not observe necrosis of the skin, muscle, connective tissue, or bone.

Discussion

Soft tissue sarcomas occur infrequently. Distinct features of these tumors are the variation in the course of the disease and the diversity of their histological forms [3]. Multiple recurrences (even in the operated region) and distant metastases are factors in determining the slow and capricious course characteristic of soft tissue sarcomas. For operable tumors, it is an established fact that when radiation therapy or surgery are used alone there is often local recurrent disease. Thus, patients should be submitted to initial extensive surgery (or systematic reoperation for initial incomplete or imprecise surgery) followed by radiation therapy (50-60 Gy) covering the excision area [8]. For this type of irradiation, electron beam therapy is used either alone or in combination with other radiation (cobalt or high energy photons).

Electron beam treatment is particularly interesting in soft tissue sarcomas from one ballistic standpoint. These tumors are located in the trunk or limbs; their volume varies from the cutaneous skin to a depth of up to 10 cm. We have seen that this type of irradiation is well tolerated when daily doses are equal to or less than 2 Gy. When the total dose is more than 60 Gy, it is necessary to combine photons with the electron beam.

The loco-regional control of soft tissue sarcomas can be achieved by combining surgery and radiation therapy. For the past 2 years our patients have been submitted to chemotherapy in a randomized trial (cyclical polychemotherapy associating Cyclophosphamide, Vincristine, Adriamycine, and DTIC) in order to decrease the incidence of distant metastases. Radiation therapy, particularly electron beam therapy (associated or not with chemotherapy) is also a palliative treatment for patients with inoperable or very advanced tumors. On the whole, these rare tumors have varied characteristics and are often poorly identified and poorly classified. Their management could be improved by:

Developing a clinical TNM staging system as proposed by the American Joint Committee [1].

Adding the grade of malignancy which also has a prognostic role.

A better coordination between surgical, radiotherapeutic, and chemotherapeutic treatment.

Comparison of results on a larger scale so that a better criticism of therapeutic conditions could be obtained.

Appendix

A. Zuppinger

It is very probable that soft tissue sarcomas are much better influenced by electrons than by photons. The irradiation should be applied postoperatively in all cases. One should not wait until a recurrence proves the real clinical malignant character of the illness because, in the latter case, the probability of distant metastases increases considerably. I should like to point out the results in a special kind of sarcoma, the *synoviomas*. This tumor was considered to be radioresistant until Sir Stanford Cade showed that amputation can be avoided in most cases by a radical excision combined with postoperative irradiation. He achieved a 5-year cure rate of 50%. We have treated seven patients with high speed electrons: six after local excision, four with clinical signs of recurrence. All these six cases have been free of disease for at least 7 years. We lost only one patient in whom an incision had been performed because of the false diagnosis of an inflammatory process. The patient was sent for irradiation a month after the histological diagnosis had been made. The tumor disappeared locally entirely, but the patient developed lung metastases. Electron therapy seems to be the treatment of choice of synoviomas and should be applied immediately after surgical excision in operable cases, or as preoperative irradiation in inoperable patients.

References

1 Cantin J, McNeer GP, Chu FC, Booher RJ (1968) The problem of local recurrence after treatment of soft tissue sarcoma. Ann Surg 168:47-53
2 Del Regato J (1963) Radiotherapy of soft-tissue sarcomas. J amer med Ass 185: 216-218
3 Enzinger FM, Lattes R, Tarloni H (1969) Histologic typing of soft tissue tumors. WHO General
4 Fletcher GH (1973) Textbook of radiotherapy. Lea & Febinger, Philadelphia
5 Hellriegel W (1976) Weichteil-Sarkome. In: Zuppinger A (ed) Symposium on high-energy electrons. Springer, Berlin Heidelberg New York pp 283-388
6 Moberger G, Milsonne U, Friberg S (1968) Synovial sarcoma. Acta Orthop Scand (Suppl) 111
7 Pritchard RJ, Soule EH, Tayler WF, Ivins JC (1974) Fibrosarcoma — A clinico-pathologic and statistical study of 199 tumors of soft tissue of the extremities and trunk. Cancer 33:888-897
8 Russel WO et al. (1977) A clinical and pathological staging system for soft tissue sarcomas. Cancer 40:1562-1570
9 Suit HD, Russel WO, Martin RG (1975) Sarcoma of soft tissue: clinical histopathological parameters and response to treatment. Cancer 35:1478-1483
10 Tapley NV du (1976) Clinical applications of the electron beam. J. Wiley, New York London Sidney Toronto
11 Zuppinger A, Poretti G (1965) Symposium on high-energy electrons, Montreux (Switzerland) 7th to 11th September 1964. Springer, Berlin Heidelberg New York

Surgical Results After Preoperative Radiotherapy of Tumors of the Head and Neck

M. Neiger

The following goals were set for the preoperative irradiation: (1) reduction of the tumor volume or achievement of operability of large tumors, (2) prevention of lymphogenous or hematogenous spread. An increase in postoperative complications, such as suture insufficiency, fistula, and arterial bleeding are the disadvantages associated with the positive effects of preoperative irradiation. The dose as well as the localization of the irradiation play an important role in the advantages and the disadvantages. A study of the literature gives no clear answer to this question, since the results of many authors are contradictory. The necessary dose is given as between 3000 and 6000 rads, with a clear increase in the operation risks between 4000 and 6000 rads. This "grey" region between 4000 and 6000 rads needs further research, on the one hand by radiobiology, on the other hand by clinical trials.

In the field of preoperative irradiation we can refer to the paper of Campana (1964) at the Montreux Symposium. In head and neck disease we would like to call attention to two points:

1) In neck dissection we observed some but not serious difficulties in the operative procedure and a few cases of delayed wound healing when the dose was around 6000 rads, which were applied only when fixed nodes were present at the beginning.

2) In preoperative irradiation of supraglottic cancers we limited the dose to a maximal 4000 rads. In a few cases mucous membrane suture insufficiency occurred which generally healed and only exceptionally needed plastic surgery. Until now we have not observed lethal severe complications.

The irradiation and the tumor size must be described with internationally comparable values. An absolute necessity in those patients treated both surgically and radiologically is a close collaboration between the radiotherapist and the surgeon, who should see the patient together from the very beginning and discuss the treatment procedure in detail.

Reference

Campana L (1964) Die Vorbestrahlung mit schnellen Elektronen. In: Zuppinger A, Poretti G (eds) Symposium on high-energy electrons, Montreux 1964. Springer, Berlin Heidelberg New York, pp 402-406

Treatment of Recurrences After Irradiation or Operation plus Irradiation

R. Greiner, P. Veraguth, and A. Zuppinger

The treatment of tumor recurrences after irradiation alone and after operation and post- or preoperative radiation is one of the most difficult problems that radiologists encounter. It is obvious that in these situations surgery should be considered first. The general experience shows that the chance of surgery is limited. With the exception of small localized recurrences, the operative procedure is often mutilating and has, in many patients, only a small probability of cure and is associated with a high risk. In addition, many patients are either very old or their general condition is so poor that surgery cannot be considered. Finally, many patients refuse extensive and mutilating surgery.

These different facts easily explain why many patients cannot be considered for surgery, and the radiologist has to decide if he will give the patient a chance of cure or palliation. One must understand that the majority of the radiologists are rather reluctant and restrict the treatment to only a limited number of patients since the chance of cure is small and the complication rate is relatively high. Furthermore, failure of control of the tumor is often interpreted as radiation damage, not only by the patient and his relatives, but also by medical doctors, with the consequence that fewer patients with good chances for cure are sent for radiotherapy. In Berne, we considered these points, but as we were not satisfied at all with the results with photons in the patients with recurrences, we tried electron therapy. As we had some good results, which we reported in 1964 in Montreux, we continued this treatment, and we think that it would be of interest for other radiologists and also for surgeons to see what can be achieved with this method of treatment, and to get some idea about how the treatment should be given and where we have to pay special attention.

We will discuss the problem in three different tumor localizations with different prognoses.

Lip Cancer Recurrences

These prove to be good indications for electron therapy. Twenty-four patients were referred in whom the extension of the tumor corresponded to T_3 and T_4 of the UICC classification of patients without previous treatment. It is very well known that these patients have a bad prognosis. Even in patients without previous irradiation Jörgensen and Elbrud had 25% recurrences after radium treatment and recommend these cases for surgery. We carried out our irradiation in such patients so that we irradiated only the tissue which would have to be removed in any cases by surgery. In an eventual later operation the loss of tissue would not be more extensive than with the primary surgery.

The lymph nodes should have prophylactic irradiation, but in some patients it was omitted. In four patients we gave preoperative treatment with doses of 3000-4000 rads at an interval of 4-6 weeks, but in two patients the tumor recurred. One has since then died, and in the other the tumor was controlled with a new electron therapy. As the possibility of cure with curative treatment was good, we abandoned preoperative treatment in lip cancer recurrences. Another failure happened in an extensive treatment in an 81-year-old patient with split therapy; however, here the dose was probably too low.

The third failure in this series was caused by neglecting the rule that these tumors should have prophylactic treatment of the submandibular region and the carotid triangle in patients with palpable enlarged nodes, which have to be considered as highly suspicious or even clinically certain to have tumor involvement. The patient interrupted the intended preoperative treatment of the metastasis and returned 7 months later with an incurable situation. In one patient only palliative treatment could be applied because of his advanced age (91 years) and his poor general health. We thus have only 3 failures in 23 patients of these series of very unfavorable cases. In addition, we treated three patients after questionable radical surgery in whom the tumor was locally controlled, but one patient died of intercurrent disease after 3 years. The dose applied varied in curative treatment between 5050 rads in 30 days, and 7960 rads in 37 days, or 1500-2330 ret. Six patients received split therapy; of these there was one failure and one residual tumor which was resected by the plastic surgeon. This needed a longer time for cure but finally healed.

These good results were also a surprise to us. We think that we are entitled to say that recurrences of lip cancer after previous irradiation, even combined with surgery, should first have electron therapy, as described in the paper by Dr. Greiner (Treatment of Recurrences After Irradiation and Operation plus Irradiation.)

Recurrences in Salivary Gland Tumors

Twelve cases of previously treated patients with parotid carcinomas were referred for treatment of their recurrences. Six patients have been free of disease for more than 7 years, and most of the others received a good palliative effect. These results are equal to those in patients who have had no previous irradiation and a correct treatment with a dose of 6000-8000 rads. The further analysis shows some results which may be of interest since parotid tumors can be controlled more easily than has been shown before. Another important advantage of this treatment is that the facial nerve can be preserved. Two failures were due to a too low voltage. The first patient with a recurrent adenocarcinoma after operation and irradiation with 15 MeV electrons with 6000 rads 5 years previously in another institution presented with the signs of a deeply situated recurrent tumor. We tried a combined treatment with 30 MeV electron and X-rays with a total dose of 6800 rads, with a very good early result, but the tumor again recurred in the depth with facial palsy and bone destruction and lung metastases at the same time. In another patient with an adenocystic carcinoma, which had been operated twice and irradiated with an unknown dose, received 6450 rads with 15 MeV, but developed a deep situated recurrence 6 1/2 years later which led to this death. These two cases support the postulate than in malignant parotid tumors the energy should be at least 20-25 MeV when electrons are used.

In four of our patients the dose retrospectively must be considered to be too low, and in only one of these patients could the tumor be controlled, but a new recurrence was successfully operated on 7 years ago. The low doses were given to avoid radiation damage. However, no patient who received less than 6000 rads showed damage later on. There is only one patient with radiation damage, a localized necrosis of the shell of the ear. This patient had received 4000 R ^{60}Co 5 years previously and a new dose of 7200 rads 15-25 MeV over 33 days. Thus, cure by local excision was easy. We may say that doses up to 3000 rads do not alter the treatment and that even with doses up to 6000 or 6500 rads in the second treatment the risk of radiation damage seems to be acceptable.

Oral Tumors

In the two tumor situations discussed the chance of controlling recurrences in former-ly irradiated tumors was astonishingly good. However, this is not valid for all tumor localizations. We have checked all patients with tumors in the oral cavity up to date. Out of 250 patients 36 were referred because of tumor recurrences after previous treatment. Twenty-two had had only an operation, and only three of these could be controlled. Out of 14 patients who had irradiation only three could be cured. How-ever, one can state that all these patients would have died without the electron beam

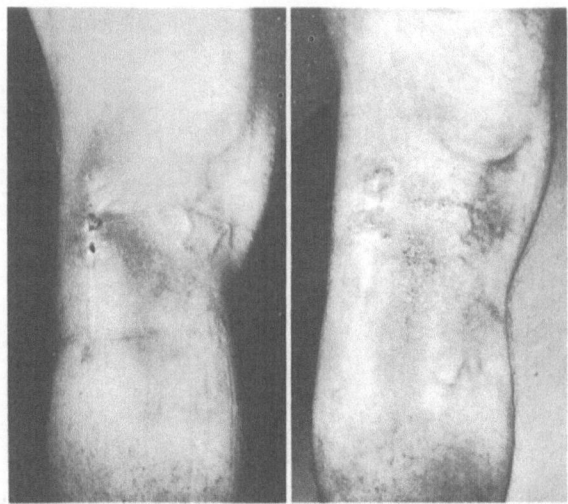

Fig. 1. 28-year-old man with an inoperable local recurrence of an osteomyxosarcoma in the bend of the left knee, 3 years after operation and conventional irradiation with 4700 rads. *Left:* before treatment; large fixed tumor mass, epilation and pig-mentation. The black points correspond to the scabs of the new biopsy. *Right:* same patient 7 years later after 22 MeV E 6880 rads, 30 days, 27 irradiations. Free of disease 20 years after second treatment

treatment and that most of those who could not be cured received a palliative effect. We see that the prognosis of recurrences of tumors of the oral cavity is poor even without previous irradiation. In our series the recurrences after operation and irradiation or irradiation alone are of the same order as the recurrences after operation alone. Here we were mainly afraid of causing a bone necrosis.

Also in other localizations surprisingly good results can be achieved. In *breast cancer* Chu has shown good local control with electrons. This is also valid for recurrences after postoperative irradiation. Exceptionally we even do not hesitate to cause radiation damage since surgery is possible. First irradiation and surgery later on has a better chance for cure than a primary operative procedure in patients with extensive recurrences. But one must be cautious in supraclavicular recurrences because of possible damage to the plexus.

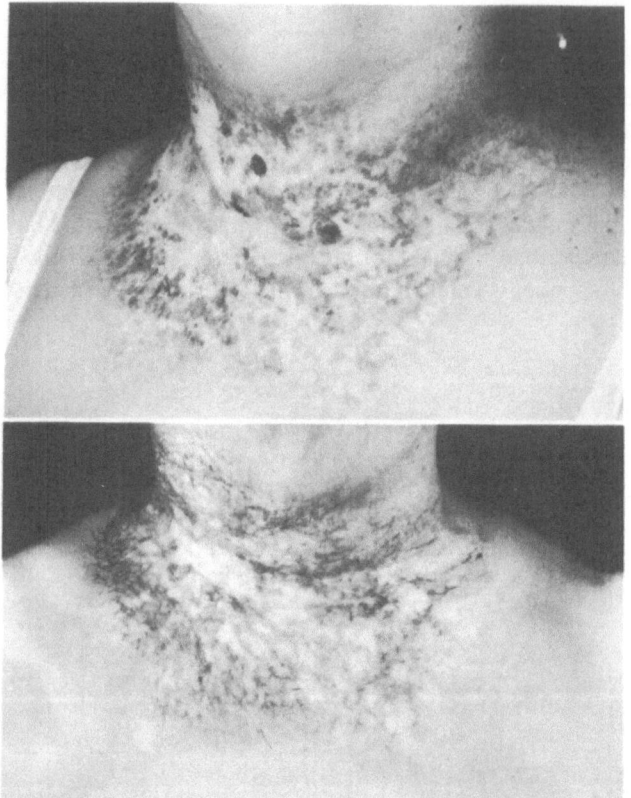

Fig. 2. 63-year-old woman with adenocarcinoma of the thyroid. Extensive recurrence after operation and irradiation with about 6000 rads (conventional radiation) 1 year previously. In spite of dyspnea, tracheostomy was refused by the surgeon because of tumor infiltration. Irradiation with 30 MeV photons 3375 rads and 20 and 15 MeV electrons 3120 rads. *Above:* neck region with severe radiation changes. *Below:* same patient 6 years later, free of disease until death (meningitis) 12 years later. Maximal additional tumor dose 6500 rads in 36 days, 25 irradiations. Dose at the surface ∼ 4000 rads

Figure 1 shows a patient with an inoperable recurrence of a soft tissue sarcoma (fibro-myxosarcoma) of the knee region after several operations and irradiation with 4700 rads X-rays. Amputation would have been the only surgical procedure. We tried the irradiation with 30 MeV electrons with a dose of more than 7000 rads. The patient has been free of disease for 20 years and fully active.

Even in patients with preexisting severe skin damage a trial may be justified. Figure 2 shows a 63-year-old woman with large local papillary thyroid carcinoma recurrence after operation and extensive postoperative irradiation. She had great difficulty in breathing and severe skin alterations. The surgeons refused tracheotomy. We had very little hope but the patient recovered. We applied a combined electron X-ray treatment with 30 MeV. She lived without great trouble, did her housework normally and died 12 years later from meningitis.

In *summary* we would like to say that electron therapy in tumor recurrences after irradiation or operation and irradiation offers in many instances surprisingly good results and should be tried in patients to whom surgery offers no good chance. We recommend applying lower single doses in the range of 150-180 rads per day, depending upon the reaction of the normal tissue and the tumor shrinkage. Also preoperative irradiation should be considered in selected patients. Though it is probable that the control of recurrences after previous irradiation is easier than with photons, we should first try to promote a good primary treatment.

References

Zuppinger A (1968) Retreatment of previously irradiated cancer electrons. Front Radiation Thor Onc, vol 2. Karger, Basel, pp 257-267
Zuppinger A, Poretti G (1965) Symposium on high-energy electrons. Springer, Berlin Heidelberg New York

Electron Treatment of Pemphigus Familiaris Chronicus Benignus (Hailey-Hailey Disease)

P.C. van der Pol

Hailey-Hailey disease is due to a genetic epidermal defect, irregularly dominant, with no sex predominance. The condition starts in adult life and tends to regress at the age of 50-60 years. The illness is located at the neck, the axillae, the inframamarian folds, the waist and the groin.

It begins with itching and formation of small vesicles which rupture and form erosions and clefts with lesions up to 15-20 cm diameter. The course of the disease is chronic, sometimes with temporary healing.

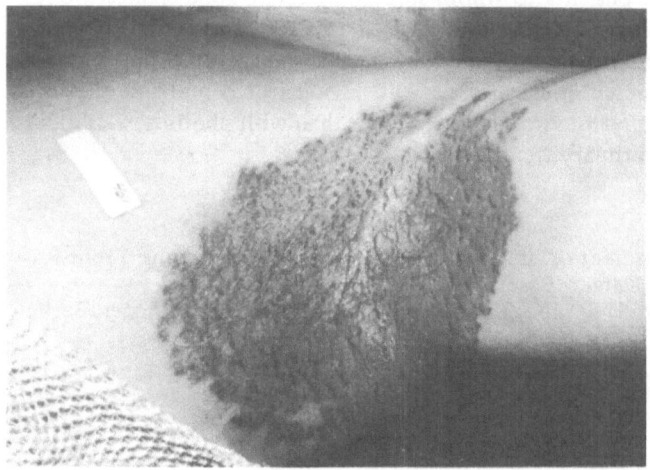

Fig. 1. Hailey-Hailey disease left axilla before treatment

Treatment with antibiotics and/or corticosteroids locally and systematic, estrogens, immunodepressants, surgical excision and autotransplantation, and local irradiation with X-rays between 10 and 80 kV with doses up to 200 R, repeated at weekly intervals, leads usually to temporary amelioration, but exceptionally to a cure.

We treated four patients with this rare illness with electrons. Three of them were cured; only one had relapses, buth she came for treatment at very irregular intervals.

We give one or two treatments a week with a total of six irradiations with 6 MeV electrons, 90 rads per session.

One patient is especially interesting: she received three series with 80 kV photons at the left axilla and one with 6 MeV electrons, whereas the other side got two series

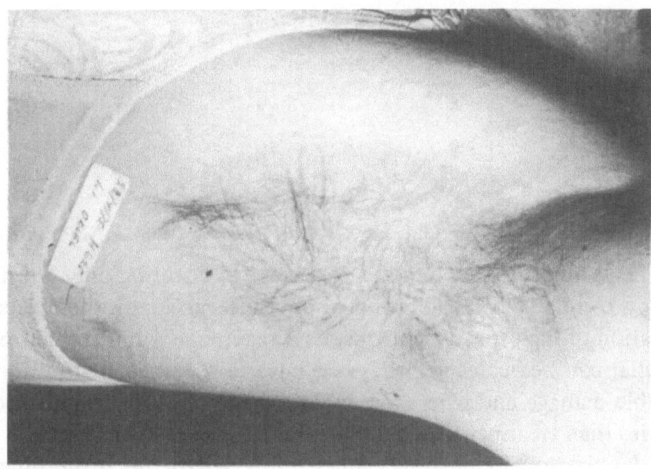

Fig. 2. Same patient 2 1/2 years later after treatment

with 6 MeV electrons and one with 80 kV photons. The overall treatment time was ten months. The side treated with more electrons is completely normal. At the left side, treated with more photons, there are fewer and thinner hairs compared with the right side. This patient is now free of disease since six years. Since this case we have abandoned X-ray treatment in favour of electron therapy.

Summary

The skin lesions of Hailey-Hailey disease (pemphigus familiaris chronicus benignus) were treated first with 80 kV X-rays, but later on with 6 MeV electrons and a shrinking field technique. Excellent results — without relapse — after treatment with electrons are shown in detail in one patient. Three other patients also responded very well to electron treatment. Two of these patients showed no relapses. One patient, who came for treatment at very irregular intervals, had a relapse.

Final Report

A. Zuppinger

This symposium has had the advantage of covering only a restricted area of radiotherapy. A relatively small group of physicists and radiotherapists, interested in the application of high speed electrons, have read papers about their experiences and discussed what can be achieved, where we have to pay special attention mainly because of possible damage and where a combination of photons and electrons offer better possibilities than treatment with either photons or electrons alone.

I have personally been very much impressed by the astonishingly good results in breast cancer and in the lymph node region of the neck. Many other facts, observations and ideas have been mentioned which may lead to progress, but where a strict proof is not yet possible. Surprising effects have been found in some tumors which have been considered to be more or less radioresistant to photons, in tumors arising in highly radiosensitive natural tissues, and in recurrences after previous irradiation. The observation that adenocarcinomas in special situations could be controlled and the effects on pancreatic tumors suggest clinical trials of preoperative treatment with electrons in deeper situated tumors, e.g., in abdominal tumors, but higher energies of 40-45 MeV are necessary.

Some speakers mentioned a better tolerance of electrons than photons so that the question arises if there is a relative biological effectiveness smaller than unity.

This version of the final report has been augmented after studying the manuscripts and including some papers of invited persons who were unable to read their papers at the symposium. The aim has been to give the reader a better survey of what can be done with this kind of treatment and where the problems lie.

I think that almost all participants of this symposium are convinced that in many tumor situations electron therapy, often in combination with photons, has to be considered as real progress, but there are still a lot of problems to be solved. I would like to suggest a collaboration in examining special questions, because very often the single therapist has too few patients to arrive at conclusions.

All those who are or have been active in the field of cancer therapy know that critical optimism, with an activity based on facts in clinical experience, and systematic work with analysis of successes and failures are indispensible for progress.

I hope that with such critical optimism and a certain tenacity we may further contribute with our tools to a step forward in the battle against cancer.

I would like to thank, also on behalf of all participants, all those who contributed to the success of this symposium by reading their papers and participating in the discussion, and especially to the Director and the organizers of the Caja de Ahoros de Guipuzcoa and to Dr. Irigaray and his co-workers and also to the BBC Company who enabled this symposium to take place by their financial support.

Collective Ion Acceleration

With contributions by C. L. Olson, U. Schumacher

1979. 63 figures, 9 tables. VII, 231 pages (Springer Tracts in Modern Physics 84) ISBN 3-540-09066-5

Frontiers in Nuclear Medicine

Editors: W. Horst, H. N. Wagner, Jr., J. W. Buchanan

1980. 198 figures, 57 tables. Approx. 350 pages ISBN 3-540-09895-X

K. zum Winkel

Nuklearmedizin

Mit einem Beitrag von J. Ammon

1975. 155 Abbildungen, 83 Tabellen. XVIII,425 Seiten (Heidelberger Taschenbuch 167) ISBN 3-540-07233-0

Handbuch der medizinischen Radiologie. – Encyclopedia of Medical Radiology

19 Bände in ca. 52 Teilbänden. Herausgeber: L. Diethelm, F. Heuck, O. Olsson, F. Strnad, H. Vieten, A. Zuppinger

15. Band

Nuklearmedizin – Nuclear Medicine

Teil 1 A:

Radiopharmaka, Gerätetechnik, Strahlenschutz. Radiopharmaceuticals, Instrumentation Technology, Radiation Protection

Redigiert von/Edited by H. Hundeshagen
1980. 244 Abbildungen in 280 Einzeldarstellungen, 65 Tabellen. 722 Seiten ISBN 3-540-08487-8

Teil 2:

Diagnostik, Therapie, Klinische Forschung. Diagnosis, Therapy, Clinical Research

Mit Beitägen zahlreicher Fachwissenschaftler.
Redigiert von/Edited by H. Hundeshagen
1978. 369 Abbildungen in 1366 Einzeldarstellungen, teilweise in Farbe, 146 Tabellen. XVIII, 1156 Seiten (51 Seiten in Englisch) ISBN 3-540-08388-X

Springer-Verlag
Berlin
Heidelberg
New York

always abreast of the latest developments in
your special field with

European Journal of Nuclear Medicine

Editor in Chief: H. Hundeshagen, Hannover, FRG

Editorial Secretary: G. Thiessen, Hannover, FRG

Board of Editors: V. Z. Agranat, Moscow;
A. M. Baptista, Lisbon; C. Beckers, Louvain;
P. Blanquet, Bordeaux; H. J. Correns, Berlin;
B. Delaloye, Lausanne; Z. Dienstbier, Prague;
A. Fieschi, Pisa; W. Finck, Rostock; L. M. Freeman,
Bronx; R. Fridrich, Basle; M. Gembicki, Poznan;
W. Graban, Warsaw; G. W. Hamilton, Seattle;
P. V. Harper, Chicago; G. J. Hine, Boulder; R. Höfer,
Vienna; W. Horst, Zurich; S. Hupka, Bratislava; Y. Ito,
Okayama; C. Kellershohn, Orsay; P. Kostamis, Athens;
J. Mallard, Aberdeen; V. R. McCready, Sutton;
T. Munkner, Copenhagen; H. Orii, Tokyo; J. Ortiz-
Berrocal, Madrid; H. W. Pabst, Munich; P. Pavoni,
Rome; G. Riccabona, Innsbruck; E. Riihimäki, Helsinki;
I. Rodé, Budapest; L. Salambashev, Sofia; K. Scheer,
Heidelberg; I. Šimonović, Zagreb; H. N. Wagner Jr.,
Baltimore; M. G. Woldring, Groningen; A. P. Wolf,
Upton

The **European Journal of Nuclear Medicine** covers the
most important developments in nuclear medicine including original articles on such vital topics as diagnosis, therapy with 'open' radionuclides, in-vitro in-
vestigations, and methods of radiobiological and radiation protection studies. In addition to a comprehensive
summary of the most significant European research
results, the journal will also carry a topical survey of all
essential data published outside Europe.

Springer
International

For subscription information and sample copy please
write to: Springer-Verlag, Journal Promotion,
P. O. Box 105280, D-6900 Heidelberg, FRG